THE CONSTRUCTION OF BUILT HERITAGE

In memory of
Pat McLernon

The Construction of Built Heritage
A north European perspective on policies, practices and outcomes

Edited by

ANGELA PHELPS
Nottingham Trent University, UK

G.J. ASHWORTH
University of Groningen, The Netherlands

BENGT O.H. JOHANSSON
University of Göteborg, Sweden

LONDON AND NEW YORK

First published 2002 by Ashgate Publishing

Reissued 2018 by Routledge
2 Park Square, Milton Park, Abingdon, Oxon OX14 4RN
711 Third Avenue, New York, NY 10017, USA

Routledge is an imprint of the Taylor & Francis Group, an informa business

Copright © Angela Phelps, G.J. Ashworth and Bengt O.H. Johansson 2002

All rights reserved. No part of this book may be reprinted or reproduced or utilised in any form or by any electronic, mechanical, or other means, now known or hereafter invented, including photocopying and recording, or in any information storage or retrieval system, without permission in writing from the publishers.

Notice:
Product or corporate names may be trademarks or registered trademarks, and are used only for identification and explanation without intent to infringe.

Publisher's Note
The publisher has gone to great lengths to ensure the quality of this reprint but points out that some imperfections in the original copies may be apparent.

Disclaimer
The publisher has made every effort to trace copyright holders and welcomes correspondence from those they have been unable to contact.

A Library of Congress record exists under LC control number: 2002102832

ISBN 13: 978-1-138-73082-3 (hbk)
ISBN 13: 978-1-138-73081-6 (pbk)
ISBN 13: 978-1-315-18811-9 (ebk)

Contents

List of Figures *viii*
List of Tables *xi*
List of Contributors *xii*
Preface *xv*
Acknowledgements *xvii*
Abbreviations *xviii*

INTRODUCTION

1 The Cultural Construction of Heritage Conservation 3
   G.J. Ashworth and Angela Phelps

PART 1: THE PRESERVATION OF THE PAST: NATIONAL CONTEXTS

2 The Conservation of the Built Environment in the UK 13
   Graham Black
3 The Conservation of the Built Environment in Sweden 29
   Bengt O.H. Johansson
4 The Conservation of the Built Environment in
   The Netherlands 45
   G.J. Ashworth

PART 2: CASE STUDIES

Theme 1: Heritage, Identity and Urban Regeneration 57
   Ingrid Holmberg
5 'Where the Past is Still Alive': Variation Over the Identity
   of Haga, in Göteborg 59
   Ingrid Holmberg

6	Nottingham Lace Market *Graham Black*	73
7	Waagstraatcomplex and Hoofdstation, Groningen: Consequence or Cause of Place Identity? *M.J. Kuipers and G.J. Ashworth*	87

Theme 2: The Heritage Site as Attraction — 101
Angela Phelps

8	Developing an Historic Monument: Reinventing the *Villa Rustica* of Gunnebo *Inger Ernstsson and Bengt O.H. Johansson*	103
9	Adaptive Re-use of Historic Properties: Wollaton Hall and Park, Nottingham *Angela Phelps*	117
10	Managing the Heritage of Fortress Towns: the Cases of Naarden and Bourtange *G.J. Ashworth and M.J. Kuipers*	131

Theme 3: Heritage as a Strategic Policy Option — 145
Bengt O.H. Johansson

11	Bolsover – after 'King Coal' *Graham Black*	149
12	Heritage in Economic Regeneration: the Case of Nieuweschans *G.J. Ashworth*	163
13	Forsvik's Bruk: A Tragic Industrial Closure or an Industrial Historical Success? *Lars Bergström*	175

Theme 4: Heritage and the Restructuring of Symbolic Places — 189
Bengt O.H. Johansson

14	Liverpool and the Heritage of the Slave Trade *Pat McLernon and Sue Griffiths*	191
15	Layers of Meaning, Layers of Space: City Strolling and the Museum Gaze *Michael Landzelius*	207
16	Shaping Symbolic Space: Parliament Square, London as a Sacred Site *Stuart Burch*	223

| 17 | Folkingestraat, Groningen: the Heritage of the Jewish Ghetto
G.J. Ashworth | 237 |

CONCLUSION

| 18 | The Experience of Heritage Conservation: Outcomes and Futures
G.J. Ashworth | 247 |

Subject Index   *259*
Place Index   *263*

List of Figures

3.1	Olaus Rudbeck dissecting geography	31
3.2	In bygone days the Goths used to drink from horns	34
4.1	Types of monuments and numbers inspected	49
4.2	Numbers of monuments in towns	50
4.3	Numbers of monuments in provinces	50
6.1	The Lace Market Area, Nottingham	74
7.1	The number of officially designated national monuments in the municipality of Groningen (1990-2000)	92
7.2	Preference of the respondents	94
7.3	Adolfo Natalini's Waagtstraatcomplex with the integrated historic Goudkantoor	95
7.4	The Groninger Hoofdstation and a part of the Post Office building seen from the Werkmanbrug	98
8.1	Villa Gunnebo with reconstructed garden	103
8.2	Plan of Gunnebo reconstructed	106
8.3	Servants' quarter at Gunnebo	111
8.4	Servants' quarter at Gunnebo, reconstruction	112

List of Figures

9.1	South façade of Wollaton Hall showing part of formal garden with original pond	117
9.2	West façade of Wollaton Hall showing nineteenth-century extension for servants' quarters and Courtyard buildings	122
10.1	Design for the rebuilding of the fortress of Naarden in 1679	134
10.2	Aerial view Bourtange 1951	139
11.1	Bolsover town centre	152
12.1	The location of Nieuweschans and other fortress towns discussed in Chapter 10	165
12.2	Modern Nieuweschans superimposed on previous fortress	172
13.1	The Göta Canal	177
13.2	A view of Forsvik's Bruk 1930	178
13.3	Kalorifer – Air Hot Machine from 1877 The situation in 1983	180
13.4	A view of Forsvik's Bruk 2001	184
13.5	Employees 1946	185
13.6	Employees in the area 2001	185
15.1	Map of Vänersborg	208
15.2	Vänersborg Museum	213
15.3	Villa Häbler, Dresden	214
15.4	Villa Häbler, Dresden, ground floor plan	217

15.5	Vänersborg Museum, ground floor plan	218
16.1	Parliament Square, London	227
17.1	Location of Folkingestraat, Groningen	240
17.2	Jewish households in Groningen 1900	241

List of Tables

13.1　Employees at Forsvik's Bruk　　　　　　　　　　　179

List of Contributors

G.J. Ashworth was educated at the Universities of Cambridge, Reading and London and has taught at the Universities of Wales, Portsmouth and since 1979 Groningen, The Netherlands. He was appointed 'Professor of Heritage Management and Urban Tourism' in the Department of Planning, Faculty of Spatial Science, in 1993. Research and publications have focused upon heritage planning and management, cultural tourism, place marketing and city centre management.

Lars Bergström whose academic background is in the conservation of built environments, has worked as the Director of Forsviks Industrial Heritage since 1993. His work includes organising and managing courses in industrial heritage, as well as activities for the increasing number of tourists who are visiting the museum. He works closely with Göteborg University and Chalmers University of Technology. His publications are mainly on the conservation and heritage of built environment.

Graham Black combines his career as a consultant Heritage Interpreter with academic work as a Senior Lecturer in the Department of International Studies at the Nottingham Trent University. His university research supports his consultancy, particularly on the use of interpretative principles to engage and involve a wider audience both in museum exhibitions and within the historic built environment. He has had a specific interest in Nottingham's Lace Market since 1978. He has published a range of articles, book chapters and book sections relating to these issues and is currently writing a book on 'Quality and Museum Presentation'.

Stuart Burch is a full-time lecturer in the Department of International Studies at Nottingham Trent University where he is currently writing-up a doctoral thesis on the history of Parliament Square. He has degrees in the History of Art from Leicester University and Sculpture Studies from Leeds University. He has studied at the Universities of Turku and Helsinki in Finland and at the Swedish Institute Stockholm University. His interests include fine art, sculpture and public monuments.

Inger Ernstsson, whose academic background is in Archaeology, Art History and Cultural Studies, has been working with cultural heritage management during the last twenty years in different positions. From 1995-2000 as a curator of Gunnebo House and Gardens, and as a developer of the cultural heritage project 'Gunnebo back to eighteenth century'. Her publications include reviews and articles on garden history.

Sue Griffiths, whose academic background is in Literature and Cultural Studies, teaches courses in Literature and contributes to the Accreditation Programme at the Open University. She has also taught at the Nottingham Trent University. Her publications include reviews and articles on literature and education. Born and brought up on Merseyside, she introduced her friend, Pat McLernon to Liverpool and its museums. She has revised and edited Pat's early draft of her chapter here, to enable its inclusion in the book.

Ingrid Holmberg, whose academic background is in the conservation of built environments, holds a Doctoral Fellowship and is member of staff at the Institute of Conservation, Göteborg University where she has taught courses since 1992. Her main interest, and subject for ongoing PhD thesis, is in the uses of the past within conservation. She currently runs a research project regarding twentieth century Swedish farmhouses and has several publications inter alia concerning the history of gardens and parks.

Bengt O.H. Johansson served in many senior positions in Swedish heritage management and was until recently special advisor to the Swedish Minister of Culture in the formulation of the Government's policies for architecture and design and for heritage management. He is the author of several books and articles on architecture and heritage (in English: Tallum, Gunnar Asplund's and Sigurd Lewerentz Woodland Cemetery in Stockholm, 1996). Johansson is presently adjoint professor in Conservation at Göteborg University and is engaged in heritage rescue work abroad as president of the Swedish Foundation for Cultural Heritage without Borders.

M.J. Kuipers finished her study, 'Urban and Regional Planning, with a specialization Heritage, Culture and Tourism', at the Faculty of Spatial Sciences of the University of Groningen in October 1999. She continued as a doctoral student at the same Faculty, under the supervision of Professor G.J. Ashworth. Her PhD research is about 'Living in the past: the residential function in the conserved built environment'.

Michael Landzelius is a Visiting Scholar in the Department of Geography at Cambridge University. His research focuses upon the politics of space, bodies and spatial practices, in the continued re-formation of modernity. He received his Ph.D. in Conservation of Built Environments from Göteborg University and was, during his post-graduate studies, a visiting doctoral researcher at the Geography Departments of Syracuse University and the University of California at Berkeley.

Pat McLernon whose academic background was in Cultural Studies and Art History, was a Senior Lecturer in the Department of International Studies at the Nottingham Trent University until her untimely death in December 2000. Her main research interest was in constructions and sites of public memory in relation to the two World Wars. At the time of her death she was one of the team engaged in researching and cataloguing memorials in Nottinghamshire as part of the Public Monuments and Sculpture Association's national project to document and record all free-standing memorials in the country.

Angela Phelps is a principal lecturer in the Department of International Studies at the Nottingham Trent University. She teaches courses in Geography and Heritage Studies with particular interests in the geography of recreation and tourism, and the recognition and management of heritage value in rural landscapes. She has research publications in heritage and tourism and has worked as a consultant in visitor management for local and national heritage attractions.

Preface

This book is the principal output of cooperation between staff of The Nottingham Trent University, Heritage Studies Division, the Department of Conservation, Göteborg University and the Faculty of Spatial Science University of Groningen, under the European Union's commission Socrates programme for advanced curriculum development. Along with other materials it is designed to support a module for the advanced study of heritage management in Europe.

The Geography Division at Nottingham Trent University, and the Faculty of Spatial Science at Groningen had cooperated successfully for a number of years in the framework of the European Union commission Erasmus and Tempus programmes as members of wider networks. I was involved in one such cooperation amongst Geographers that led to the publication of the *Management of Urban Change in Europe* (1997) edited by Alan Dingsdale from Nottingham and Paul Van Steen of Groningen. An emerging interest in heritage management then produced the idea of cooperation in this area of studies.

When the EU commission transformed its educational programmes under the new umbrella of Socrates the opportunity arose to cooperate on the preparation of an advanced curriculum programme. It was decided that an academic course in heritage management would offer an attractive project to support vocational and professional aspects of the subject. It was felt at Nottingham and Groningen that the participation of an institution whose members had specialist knowledge and skills in the process of conservation, to compliment their own geographical perspective, would be desirable. It was also felt that an institution from another EU member state with whom the existing partners had not previously cooperated would also be an advantage.

Enquiries among heritage experts who know the European universities well suggested that the Institution of Conservation at Göteborg would be the ideal partner. So it was that the three partners came together and embarked on a three-year project to produce innovative academic and curriculum ideas in the area of heritage management. It was also, in a sense, where my contribution ended, because I have some skill in curriculum design and project management but little knowledge of heritage management. However, with the extraordinary goodwill of the experts I was able to find a minor role outside the academic aspects of the project to enable me to enjoy with the others stimulating and convivial experiences in Göteborg, Nottingham and Groningen.

The resulting work has produced a series of illuminating case studies. The team members were convinced that the shared experience demonstrated by these cases takes the study well beyond the geographical boundaries of the countries represented by the partner institutions. Consequently this book has been prepared to draw common themes together and address the urgent questions relating to the consequences of heritage conservation practice. The work shows that while the process of heritage conservation was largely worked out through the twentieth century, the more challenging project of heritage management will be a continuing challenge for the twenty-first century.

Alan Dingsdale
Nottingham Trent University
December, 2001

Acknowledgements

Many people have assisted the authors in the preparation of this book. The Socrates project team is particularly indebted to Charlotta Hanner-Nordstrand from the Institute of Conservation in Göteborg and Alan Dingsdale from The Nottingham Trent University for essential administrative support and pertinent insight into the selection and development of the case studies. The authors have enjoyed fruitful exchange visits in the course of which we have been able to discuss heritage issues while touring different examples of heritage management solutions. Our particular thanks go to the managers and staff who took time to welcome us to Forsvik Bruk, Gunnebo, Wollaton Hall, Rufford Country Park, Newstead Abbey, Nieuweschans Spa, Göteborg City Planning Department and the Mayor of Gemeente Riederland. Linda Dawes drew the original maps and diagrams, and Phil Pierce scanned some of the illustrations. Olwyn Ince took on the onerous task of preparing the text supported by Janet Elkington and Robin Conway at The Nottingham Trent University, with most helpful advice from Ruth Peters at Ashgate.

Abbreviations

DoCoMoMo	International Group for the Documentation of the Modern Movement
DofE	Department of the Environment
ERDF	European Regional Development Fund
EU	European Union
GKH	Göteborgs Kommunfullmåktiges Hardlingar
ICOMOS	International Council on Monuments and Sites
MORI	Market and Opinion Research International
PPG	Planning Policy Guidance
RCHM	Royal Commission on Historical Monuments (now part of English Heritage)
SOU	Statens Offentliga Utredningar
SRB	Single Regeneration Budget
UNESCO	United Nations Education, Scientific and Cultural Organisation
VAT	Value Added Tax

INTRODUCTION

1 The Cultural Construction of Heritage Conservation

G.J. ASHWORTH and ANGELA PHELPS

The turn of a century prompts reflection on the past and speculation about the future. The late nineteenth century saw a remarkable rise in interest in aspects of preservation, recognised in the flowering of a museum culture throughout Europe and a new focus on both the built environment and the 'natural' landscape. The shift of interest resulted in relicts of past eras being collected, not just as curiosities or for aesthetic pleasure, but as testimony and object of study. The late twentieth century also witnessed a revival of interest in the past, but in a form that prompted criticism for moribund nostalgia. Some aspects of the 'heritage industry' do represent an obsessive commodification, but there is growing recognition that positive assimilation of the past may provide an essential cement to hold together an increasingly disparate and dissonant present.

The central thesis of this book is that the heritage of the built environment is not a result of haphazard survival, but rather the outcome of individual and group consciousness relating to a 'sense of place'. The built environment as it has survived is a cultural construction, its appearance and meanings dependant on a complex process of selection, protection and intervention. Nevertheless, it is recognised that much of this selection has been subconscious, or dependent on individual obsession rather than coherent overview. The contribution of the twentieth century has been a gradual revealing of a consciousness for conservation, and its incorporation into legal frameworks prioritising public, over private, interests. Despite the persistence of private property ownership, there is increasing expectation in both town and countryside of a public 'ownership' of, and access to, the landscape. The challenge for the twenty-first century will be to integrate the conservation project more effectively with future planning and deal with the issues resulting from the domination by selected public interests. An appreciation of past achievements should not result in efforts to preserve past forms as empty relics, but to conserve evidence within vibrant communities that are forward looking.

The Evolving Forms

The built environment evolves through a sequence of building, renovation, demolition and rebuilding. Prestige buildings have been protected by their purpose and inspirational values for centuries. However, the deliberate selection for retention of buildings as exemplars of type or style is a recent phenomenon. In the course of the twentieth century this process has passed from private hands to public; it has become marked by the organised intervention of the state, and in some cases by international organisations such as UNESCO. Whilst the main purpose of planners has been the efficient use of land and management of traffic, particularly in pressured urban areas, another project has gradually emerged: the deliberate conservation of buildings from the past.

The purpose of this book is to chart the progress of the conservation project in relation to the built environment. The case studies illustrate how this process has progressed from haphazard to more deliberate intervention, hence the idea of 'constructing' a heritage. Two themes run throughout these papers: the process by which certain buildings, and complexes of buildings, have become valued above others and selected for conservation; and the means by which conservation has become embedded within legislation. The studies demonstrate a range of outcomes, some deemed successes, others failures; the cases also demonstrate that the criteria used to judge success and failure are neither consistent nor stable. Despite this, common outcomes recur, suggesting the emergence of what may be declared a heritage project of the late twentieth century. The following studies offer an exploration of the development of conservation, based on a comparative analysis of three countries. As such it provides insight into the experience of different localities. However the purpose of the authors goes further than documenting a process of change and recovery. What emerges questions the relevance of national perspectives in explaining the heritage project.

The Study Area: Legal Frameworks

The first block of chapters considers the national contexts for conservation of the built environment and consists of three chapters each describing a single national case. This requires two major initial justifications: first, the prominent choice of the national spatial scale and secondly, the selection of the three specific cases from among the many possible.

The first justification is simply that there has been, in the European historical experience an intimate relationship between the creation of heritage, its legal, organisational, and management structures, on the one side and the political notion of the nation-state. The discovery and propagation of a national heritage was an essential precondition for the establishment of the nation-state. However the nation-state, once established, needed to legitimate itself through the creation and propagation of a national heritage. The result is that national governments have in the last 150 years taken an increasingly important role in the identification and presentation of heritage in many of its forms. The cases below describe how, over a period of around a century and a half, the role of national governments throughout Europe, has assumed an ever increasing responsibility for most aspects of heritage. The point has now been reached in many fields of culture and the arts where the national government exercises what amounts to a monopoly of legal and financial controls and conversely many aspects of heritage are now almost completely dependent upon national designation, funding and agency management.

All three national cases have a millennium or more of recorded history of settlement, have experienced most of the important political, cultural and economic movements and revolutions of this most restless of continents and all three have at various periods occupied central positions on the European or even global stage. Thus they demonstrate such major continent wide themes in the history of heritage conservation as the political struggle between liberal and social democratic values, the economic dilemma of change through development or stagnation through preservation and the cultural choices available between a clear and unambiguous homogeneous national heritage and the rich diversity of a potentially fissiparous multi-religious, multi-cultural and multi-racial heterogeneous heritage.

The UK is the most instructive case, if only because it is one of the larger political entities in Europe. Equally however its inherent multinational structure illustrates the changing patterns of tensions and reconciliations between the heritages of various social and cultural groups. Further, early industrialisation and subsequently early de-industrialisation has presented the problems and possibilities of a post-industrial, service dominated, leisure orientated society in which historicity in various forms has become of increasing importance. It is not surprising therefore that in many aspects of heritage interpretation and management and in heritage and cultural tourism, the UK is the European market leader. Finally the global economic and political role exercised by Britain over two centuries has both spread aspects of British culture, and thus its heritage, around the world and conversely brought many different cultures and their heritage to

Britain. The first phenomenon gives British heritage a world importance and market, while the second creates the difficulties and opportunities of multi-ethnic heritage within Britain.

In sharp contrast to the UK, the other two cases are countries with small populations and a smaller impact upon European or world affairs. Nevertheless both have had extensive overseas and especially trans-Atlantic connections and both have strong social democratic political traditions which have resulted in the establishment of effective land-use planning systems and a broad based public acceptance of strong government intervention, not least in the field of heritage.

Sweden combines a small population with an extensive and in part sparsely populated land area and is in this respect the mirror image of The Netherlands. Government involvement in heritage was originally motivated by a perceived need to reinforce a concept of a distinctive Swedish identity within Scandinavia, especially in contrast to the previously politically dominant Danish heritage. This concept of Sweden was even projected onto a wider European stage through the idea of a 'Gothic' contribution to European civilisation. More recently other motives and identities have been added, whether these are on the scale of Scandinavia, 'Norden', or even Europe and the world or at the other extreme more locally varied heritages of the cities and the regions.

The single most relevant attribute of The Netherlands is its small physical extent relative to its population. This has resulted in a population density and compactness of settlement form which presents special challenges but also accounts for the well-developed spatial management system. A relatively homogeneous national culture, society and government has since the Second World War been substantially modified by post-colonial immigration, more recent labour immigration from Southern Europe and North Africa, and more generally the globalisation of many cultural ideas. The operation of these factors explains some of the current tensions in Dutch heritage interpretation. In particular there is the contrast between, on the one hand, the projection of a homogeneous national image from an unambiguously interpreted past and an emerging more regionally, ideologically, ethnically and politically nuanced heritage that reflects more closely many of the more complex realities of the country.

The three national cases are presented in this section of the book for comparison and any comparative approach involves considering simultaneously the two contradictory attributes of similarity and difference. The assumption is made here that there are sufficient common features in the experience of the three countries to make a comparison possible but equally that enough differences emerge to render it worthwhile.

The common features sought are reflected in the sub-divisions of each chapter. A broadly chronological sequence is followed, beginning with *awakening of public opinion* or at least that section of opinion that exercised a leadership in taste-forming and ultimately decision-making. It is notable that although each country produced its own prophets and campaigners, there was considerable international interaction between them in a sort of European platform favouring building preservation. Success in arousing the commitment of a consensus of the informed resulted, however falteringly, in *the establishment of national frameworks* for the identification, inventorisation, grading, legal protection, commitment to maintenance and even support for restoration and rehabilitation. *The operation of the systems* led to a confrontation of practice with many technical, administrative, political or philosophical difficulties. Finally, although it may be premature to declare victory in the crusade begun a century and a half ago to preserve the preservable past, it is clear that the problems faced today are quite different from those faced by the crusaders. It is *the consequences of success* that should now be the focus of the attention of heritage planners and managers.

Although each national case follows this sequence, and there is considerable congruence detectable, the details of the timing and of the nature of the legislation and organisation vary according to the political and ideological cultures of the countries concerned. It is these variations that can be identified within such a comparative context that provide some of the most interesting themes and approaches evident in the remaining sections of this book.

Constructing a Sense of Place: Thematic Studies

The case studies have been selected to exploit the opportunities of comparative study, by exploring common themes within the national perspectives of the three chosen countries. The under-pinning concept is the relevance of a 'sense of place', which consolidates the meanings and values of different communities with the environments in which they arose. What emerges from these studies is a reinforcement of the values that have become labelled as 'heritage'. These concern the recognition of the importance of retaining buildings of significance, not just for architectural interest, but as markers within an increasingly complex and fast-changing world. The studies explore the way in which significance is determined, emphasising the importance of location, cultural context and social use rather than scale. Although some examples demonstrate an easy placing of heritage importance on single buildings of great note, significance is also found for more modest buildings that reflect through their use or composition strong culture and social meanings.

The case studies have been grouped according to prominent themes to assist comparative analysis. The first theme considers the role of heritage in place identity in the urban environment. In the late twentieth century many cities contained run down areas where the prime industry had declined. Urban regeneration has been pursued using a variety of vehicles, but all in some way hinge on heritage meanings. In some cases heritage stories have been used to promote a revitalisation based on visitor attractions, in other cases 'heritage' features have been used to retain a sense of identity in otherwise totally refigured locations. Other examples show how places may become soulless if their past is ignored. Nevertheless, the studies question the assumption that retention of past meanings will enrich current community life. There is a risk that the heritage project becomes a means of financing the refurbishment of the fabric, but detached from the culture that gave it substance. A key issue relates to the current and future users of the areas. There is a tension between the recovery of a locale for its residents and the creation of an attractive area for visitors. Relating heritage stories to help visitors understand the past importance of an area may actually alienate current residents.

The second theme concerns the development of heritage sites as visitor attractions. A recurrent theme throughout the studies is the need to find viable uses for otherwise redundant buildings. Conservation cannot succeed without new uses being found that integrate the retained structures within a developing community. This section considers examples where there has been a deliberate choice to develop a visitor attraction as an economic activity, to generate income sufficient to maintain the conservation project. The current interest in heritage tourism may well provide opportunities, but the need to make such options sustainable and self-financing moves the priority away from building conservation or heritage interpretation, to income generation. This creates new pressures that may undermine the overall purpose.

The third theme looks at the potential of heritage as a strategic policy option. This section addresses the difficulties of re-development more directly, looking at areas that have lost their prime function. Can heritage be used to create a sustainable future? The studies challenge the gloss of heritage interpretation and recognise the importance of involving the community in a process of self-discovery. However, heritage re-built in this way may differ significantly from the past that the projects were devised to protect. Which case takes priority?

The final theme deals with the difficult issues arising out of the symbolic restructuring of places associated with heritage meanings. Perhaps the most controversial of all, this section deals with highly emotive issues where meaning is contested and dissonance undermines unified projects. It would be easy to find current examples of the exploitation of heritage in fuelling both internal and inter-national conflict in the world today. What makes these studies of particular interest is the recognition of dissonance within seemingly peaceful areas. They demonstrate how the tensions may be just beneath the surface. Revealing, recognising and resolving such tensions consolidates the purpose of the heritage project.

Heritage Futures

The case studies discussed in this book reflect research completed over the last five years. They reflect the vibrancy of conservation work, the range of buildings now found worthy of retention and the remarkable ingenuity of the people engaged in the practice. Nevertheless, events evolving while the research was concluding throw a dark shadow over such endeavours. The act of conservation identifies buildings that are considered of heritage value, sometimes locally, often nationally. Such action should be a source of affirmation and pride. Sadly, it increasingly presents targets at time of conflict. The experience of the states of the former Yugoslavia shows dramatically how heritage values can be turned inside out to create new ways of hurting communities. The deliberate desecration of places endowed with religious beliefs may be used to mark the supremacy of a system and attacks on renowned sites such as Dubrovnik, inscribed on the World Heritage Convention list, may be used to focus international attention. However, nothing could have prepared people for the ferocity of the attack on 11th September 2001 on the World Trade Centre in New York. Minoru Yamasaki's great towers were the tallest buildings in the world when dedicated in 1973; although neither beautiful nor innovative, their indelible association with American cultural capital became their downfall, at the cost of many thousand lives.

Amongst the outpouring of condemnation and painful description, there is the chilling reflection that attitudes towards heritage values may also need to be reassessed. The loss of these buildings has abruptly changed one of the most famous city skylines in the world. When the painstaking clearance has been completed a decision will have to be made whether to attempt to rebuild in defiance of those bent on destruction, or whether a more fitting memorial is an empty shell, as in the old Coventry Cathedral destroyed by bombing in World War Two. The scale of the destruction, in the heart of a global city, renders rebuilding inevitable, but how much and in what style will engage city planners for years to come. Heritage conservation in the future will not only be a matter of conscious acts of memory, but will increasingly require courage to identify, protect and continue using the buildings that reflect community aspirations. The need for the purposeful celebration of the variety of people's heritages has never been more pressing.

PART 1
THE PRESERVATION OF THE PAST: NATIONAL CONTEXTS

2 The Conservation of the Built Environment in the UK

GRAHAM BLACK

Three characteristics are especially important in this field. First, the UK was the first country to experience the industrial, and accompanying urban, revolutions. Consequently, the radical shift in economic activities and settlement patterns was both experienced much earlier and more completely than on the continent. By 1851 more than 50% of the rapidly growing population already lived in towns and cities; by 1901 the figure was nearer 90%. It is not surprising that the reaction to this rapid change was also apparent early in the nineteenth century and that the British were in the forefront of the crusade for the preservation of aspects of the past. Secondly, the UK has been a leading player on the world economic and political stage for some centuries. More recently it has experienced an increase in the global significance of its language and associated cultural productivity. This has rendered UK heritage accessible to, and important for, a world-wide market, with evident consequences for the quantity of conserved structures and monuments, the nature of the interpretation and the size of the visitor market. The UK is thus a world leader in heritage tourism. Thirdly, and related to the UK's history of global involvement, is the growth, especially since the Second World War, of large racial and ethnic minorities which, especially in the major cities, are developing a pronounced multicultural society which again has implications for what is conserved and how it is interpreted.

A legislative and practical framework for the conservation of historic buildings and areas is now an accepted part of national policy in the UK, attracting strong popular support for what are widely recognised to be assets of immense historical, cultural and social value. A century ago, the situation was very different. While a few lone voices called for protection, the UK government was still heavily influenced by liberal traditions of individual responsibility rather than state intervention and by an overriding commitment to private property rights. To achieve today's conservation machinery, both these had to be overcome.

Issues surrounding the public's willingness to accept the expansion of State intervention in general were important to the development of conservation strategy. In particular, the issue of private property rights is central to the development of conservation policies. Legislation for the conservation of the built environment permits the State to place major restrictions on the property rights of the individual, purportedly because this reflects, 'a sense that the relics of the man-made past are important enough to inspire such appreciation and to justify such restriction of property rights in the wider interest' (Hunter, 1996:1). This represents a sea-change in opinion and reflects the interdependency of the themes discussed below. Would the cumulative development of conservation legislation and policies have been possible unless supported, or even driven, by public opinion?

The Awakening of Public Opinion

A state-based conservation strategy is a very recent phenomenon in the UK, with the first legislation only appearing in the 1880s, no extensive State action until after 1945, and the most important developments with regard to the built environment only taking place since 1967. Its origins are in the rise of the antiquarian movement and the influence of a small elite. William Camden, the founder of British archaeology, published his historical and geographical survey of the British Isles in 1586. John Aubrey (1626-1697) wrote the first British book devoted to archaeological remains and also attempted a history of medieval architecture. The topographical writings of William Stukeley (1687-1765) included extensive coverage of British antiquities. As well as writers there were collectors and cabinets or galleries to hold and display the collections. However, concern for British antiquity was limited.

A resurgence of interest in the classical past amongst the aristocracy of the sixteenth and seventeenth centuries was followed by the 'golden age' of travel and acquisition during the Grand Tours of the eighteenth century. So long as the classical ideal dominated the thinking of the elite, there was little likelihood of an appreciation for older British buildings and monuments on a scale that might influence events. There were occasional antiquarian voices, such as Richard Gough or John Carter in the late eighteenth century. However, what mattered most was a switch in taste towards 'Romanticism' and the 'Picturesque'. Initially, this was a cult of wild nature, but old buildings soon came to be valued for the harmonious way in which they merged into the landscape. A major literary influence came from writers such as Sir Walter Scott whose hugely popular historical novels, such as *Ivanhoe* (1819), encouraged a widespread fascination for all things medieval. General interest is reflected, for example, in the opening of the Tower of London to the public in 1828. By the 1850s it was attracting over 200,000 visitors a year.

Despite this emerging interest in historic buildings, particularly amongst the wealthy, educated elite, it is no surprise that the earliest legislation related to archaeological sites and monuments only. These were not lived in, so the issue of private property rights, though still vital, was not necessarily paramount. Equally, sites such as the Tower of London or Stonehenge came to be seen as 'national antiquities', associated with the rise of nationalism which was a feature of much of Europe in the nineteenth century. 'Gradually ... such feelings provided a seedbed for seeing such monuments as especially precious, and hence the State as having a role in guarding them, rather than leaving them to private individuals' (Hunter, 1996: 5).

The Ancient Monuments Protection Act 1882 is the first conservation law in the UK. It followed proposals made since the 1840s and eight failed attempts from 1873-1880 to introduce more extensive legislation to protect archaeological sites. These faced overwhelming opposition in Parliament because of concerns for interference with private property rights and the potential cost to the taxpayer. It is equally unsurprising that this first UK Act relied on voluntary initiatives and the State could take sites into guardianship only with the agreement of the owner. However, the growing interest in archaeological monuments and relics did coincide with increasing attention being given to ancient buildings. We see this most clearly in the 'repair' and 'improvement' of medieval churches and cathedrals, beginning with the work of the architect James Wyatt in the late eighteenth century. Many of the 'improvements' made by Wyatt and his successors, such as Giles Gilbert Scott, were highly destructive of original detailing. Literally thousands of medieval churches were affected. The angry reaction of some scholars and antiquarians resulted in a very different view of how historic buildings should be protected. 'We have no right whatever to touch them. They are not ours. They belong partly to those who built them, and partly to all the generations of mankind who are to follow us' (Ruskin, 1880:197).

Meanwhile Romanticism began increasingly to influence new architecture with the rise of Gothic Revival, based on a strong appreciation of medieval architecture. The best known supporters of this movement are probably Augustus W. N. Pugin, architect of the Houses of Parliament, the writer and art critic John Ruskin, and the architectural theorist Eugene

Emmanuel Viollet-le-Duc. These writers were all looking to the new as well as to the historic. However, the effect of their work, and that of their supporters, in influencing public opinion towards conservation was profound: '(there could be no action to preserve monuments) unless there be in the mind of the people a certain force of intelligent belief' (Brown, 1905:31-2). What we can now see, in retrospect, is the piecemeal nature of legislative provision, linked closely to upsurges in public interest. To understand the development of conservation legislation in the twentieth century we need first to explore how public opinion was galvanised.

First, one must acknowledge the campaigning role of national amenity societies, beginning with the foundation, by William Morris, of the Society for the Protection of Ancient Buildings (SPAB) in 1877. These societies worked particularly as educators and lobbyists. Most members have been enthusiastic amateurs, with their leadership frequently journalists or architectural historians, rather than architects. They have worked ahead of fashion. Thus, the Georgian Society was founded in 1937 out of rising concern over the rate of demolition of eighteenth century buildings between the two world wars, while the Victorian Society was founded in 1958 at a time when nineteenth century British architecture remained largely unloved. The process continues. For example, DoCoMoMo was founded in 1990 to defend the products of the Modern Movement. The real influence these societies now possess is reflected in the responsibility, enshrined in legislation, that a number have to comment on relevant listed building planning consent applications.

Alongside these are societies which have believed in taking more direct action. Best known is the National Trust for England, Wales and Northern Ireland, and that for Scotland, with their policy of conservation through acquisition. The National Trust is now the largest private landowner in Britain and played a central role, for example, in the conservation of, and development of public support for, the country seats of the British aristocracy. A different example would be SAVE Britain's Heritage, an organisation which grew from the 1974 exhibition at the Victoria and Albert Museum in London, the *Destruction of the Country House*. This is much more of a ginger group, with no ordinary members, which is expert in the generation of publicity on individual threatened buildings. However, it has fully understood the need to go beyond crying wolf. For buildings to be saved, they must be given a viable new use and SAVE has made practical proposals on how this can be done.

While these societies developed on a national scale, the period since 1945 has seen the rise of a grassroots movement, the local Civic Society. Today there are over 900 Civic Societies in the United Kingdom, with c. 365,000 members. They are represented nationally by the Civic Trust, founded in 1957, but most of the work is done on a local level, from campaigning about specific development proposals to raising local awareness in the built heritage. Civic Society members make photographic records, scrutinise local planning applications, promote quality new architecture and conservation schemes, run lecture programmes and provide trail leaflets and guided tours. Societies have been effectively supported by the media. The early development of a lay interest in architecture owed much to magazines such as *Country Life*, founded in 1897, *Picture Post* and *Amateur Photographer*, alongside the books produced by publishers such as the Batsford Press and, later, the monumental *Buildings of England* series by Nikolaus Pevsner. Later, radio and, particularly, television played a key role in widening interest amongst the public at large. For example, the television appearances of the poet Sir John Betjeman were crucial in influencing public opinion in favour of Victorian architecture. The 1960s and 1970s witnessed a rise in campaigning journalism regarding the destruction of the built environment, linked to an intense public reaction against the unprecedented scale of large area redevelopment then taking place. Major campaigning books, such as Fergusson's *The Sack of Bath* (1973) and Amery and Cruikshank's *The Rape of Britain* (1974), showed the impact of comprehensive development in historic towns.

Since then, the pressure for change has taken on a new form which it has been much more difficult to galvanise public opinion against. Rather than total demolition of large areas, we have witnessed the gradual alteration of detail and the introduction of small-scale infill which together, in the long run, can have as devastating an impact. Yet, with no great battles to be fought, it has been much harder for amenity societies to generate publicity. The battle for the hearts and minds of the British public has been won, but the war with developers continues in a more insidious form. 'Most people place a high value on the historic environment — 87% think that it is right that there should be public funding to preserve it. 85% think it is important in the regeneration of our towns and cities. 77% disagree that we preserve too much. It is seen as a major contributor to the quality of life. The historic environment is seen by most people as a totality. They value places, not just a series of individual sites and buildings' (MORI survey of people's attitudes to the historic environment quoted in English Heritage, 2000:1).

Establishing the National Framework

Listed Buildings

On what basis should a State make a decision to protect in perpetuity examples of its built environment? Previous generations were happy to replace old buildings with new, normally for purely practical reasons. The new was not seen as a threat. Buildings, unlike archaeological monuments, were lived in. To ban change was a direct interference in the rights of the law-abiding property owner to do with his building as he saw fit. Nevertheless there is substantial evidence in Britain from the 1870s, of concern over the destruction of individual buildings while, from 1867, the Royal Society of Arts began fixing commemorative tablets to identify the homes of famous people. By c. 1900 there was a more widespread reaction. The National Trust was founded in 1895 (and the National Trust for Scotland in 1931). In 1896-7 the London County Council instituted mechanisms to act if an historic building was threatened. The first volume of the *Survey of London* appeared in 1900. In 1908, three Royal Commissions on Historical Monuments (RCHM) began to work, in England, Scotland and Wales. Their role included the making of an inventory, 'of the ancient and historical monuments connected with or illustrative of the contemporary culture, civilisation and condition of life of the people...from the earliest times to the year 1700' (Royal Warrant establishing the RCHMs, 1908). Lord Curzon reflected the growing public opinion in his speech during the debate on the Ancient Monuments Consolidation and Amendment Act 1913. 'They are part of the heritage of the nation, because every citizen ... feels an interest in them although he may not own them ... I believe they (the owners) do generally recognise that they stand with regard to these monuments not merely in the position of private owners of property, but that they are owners of that which is, in a sense ... a national possession, for which they are trustees to the nation at large' (Lord Curzon, Parliamentary debate 1912, quoted in Hunter, 1996:44).

While the legislation remained weak in terms of the protection afforded, the principle of protection in the public interest, including restrictions on the rights of the private owner, had been accepted, at least with regard to ancient monuments. Once this basic decision is made, how does one decide which buildings should be deemed worthy of retention and who should make those decisions? At this stage the brief to the RCHMs, referred to above, shows that what were considered important in 1908 were medieval churches, castles and other structures, together with early specimens of vernacular architecture, and culminating with the heroic age of the architect Wren.

It was not until the Town and Country Planning Act 1932, fifty years after the pioneering Ancient Monuments Protection Act, that local authorities were given power to make preservation orders for buildings of special architectural or historic interest. However, each order had to be approved by the relevant government minister. The reality is that the development of both conservation powers and criteria for selection had to depend on a combination of the overcoming of prejudices and the development of new areas of interest. Latterly, both economic justification and the selection of exemplars have taken on increased relevance.

When the first lists of key buildings were produced as an emergency measure in 1940-1942, following the first serious air raids, 'the dating criteria were to include all medieval buildings; all good examples of any category down to 1750; and from 1750-1850 only buildings outstanding in their class. In this last division, selection was to be made sparingly and with great care' (Harvey, 1993:1). The criteria for selection defined here, and in 1947-1948, reflected a prejudice in favour of individual buildings and a bias in favour of the work of specific architects, especially pre-Victorian ones. The vernacular tradition and more repetitive terraces or groups of buildings were seen as of inferior quality, whatever their attractiveness in terms of the built environment. One result was that in this first listing, whereas some of the towns were covered fairly adequately, the survey of rural areas tended to be no more than a gesture.

These initial lists were begun in 1947 and were expected to take three years. The primary function of the first list was to provide unambiguous information for town planners, and for private owners, on what buildings ought to be kept in the much-heralded renewal of Britain after the war. Their introduction was, therefore, both a protection measure and a support for re-development. An inventory of what society considered to be of value would make informed decisions possible. When the first listing was completed in 1970, twenty years late, it was universally acknowledged as inadequate: a national re-survey was already under way using much wider criteria and the concept of Conservation Areas had been introduced to provide protection for groups of buildings and the spaces between them. The re-survey, accelerated in 1979, resulted in a fourfold increase in the number of listed buildings compared with those included in the first list. It was completed by 1992-1993. This re-survey was subject to much wider selection goals as a result of: strengthening of the concept of group value; the overcoming of the widespread prejudice against Victorian architecture; a substantial rise in public interest in vernacular, industrial and farm architecture; and increased lobbying by amenity societies concerned with twentieth century buildings. The re-survey did not just reflect widened criteria, it also reflected the more detailed work now believed to be required, including an investigation of the building's history, architectural and structural characteristics and condition.

From the initial RCHM cut-off point of 1908, the single criterion for the listing process today is that the building must be at least ten years old. The definition of what constitutes a building suitable for listing has also been made as broad as that for an archaeological monument, including, 'anything from a street lamp to a railway viaduct, a horse trough to a textile mill' (Studdards & Hargreaves, 1996:2). The current criteria for listing, as defined in *Planning Policy Guidance: Planning and the Historic Environment* (DofE, PPG15 1994) are 'architectural interest', 'historic interest', 'historical association' with nationally important people or events and 'group value', especially where buildings comprise an important architectural or historic unity (DofE, PPG15, 1994:26-27). Other considerations taken into account include age and rarity. 'All buildings built before 1700 which survive in anything like their original condition are listed; and most buildings of about 1700-1840'. Additionally 'after about 1840 ... greater selection is necessary to identify the best examples. Buildings which are less than 30 years old are normally listed only if they are of outstanding quality and under threat. Buildings which are less than ten years old are not listed' (DofE, PPG15, 1994: 6.11, 27). So currently criteria of age, quality, scarcity, completeness, 'sense of place' and association all influence selection for listing. The final decision remains with officials and a government minister at national level.

Within the overall listing, the importance of individual buildings has never been seen as uniform. Percentages shown in list entries for England in 1993 are: Grade I 9,000 (2%), Grade II* 18,000 (4%) and Grade II 416,000 (94%) (DofE, PPG15, 1994: 6.6, 26). Grades I and II* are for buildings of outstanding architectural or historical interest (about 6% of all listed buildings). Grade II (about 94% of all listed buildings) is for buildings representing a major element in the historic quality of towns, villages and countryside (DofE, PPG 15, 1994: 3.6,8).

With the completion of the national re-survey, the policy of English Heritage, the body now responsible for listing in England, has switched to one of detailed research of types of buildings and the selection of exemplars. This began with a survey of cotton mills in the Manchester area which revealed that the previous approach of recommending the more architecturally elaborate buildings was missing the totality of the industrial process. There is now a switch to the selection of whole industrial sites.

This will inevitably lead to conflict as ancillary buildings could well be of inferior quality. English Heritage are well aware of the need to select for listing those sites which have a viable economic chance of survival. Alongside this attempt to refine existing criteria, there is potential for a continued widening to reflect the rise of new interests. In the last ten years the Garden History Society (established 1965) has prompted the creation of an initial Register of Parks and Gardens of Special Historic Interest, but this does not have statutory effect.

A remaining prejudice to be overcome relates to public attitudes to buildings of the Modern Movement and later. The first proposals for listing post-1945 buildings put forward by English Heritage, in 1988, saw few accepted by the Secretary of State. The policy here also is now in favour of research followed by the recommendation of exemplars. Although a limited listing of such buildings has begun, the approved lists have yet to win favour with the public.

Conservation Areas

By the mid 1960s the concept of 'group value', referred to above, had already been widened to reflect concern to protect the settings of individual historic buildings. Even the Town and Country Planning Act 1932 had referred to 'preservation schemes for buildings and groups of buildings'. However it was only with a Court of Appeal judgement in 1964, relating to the protection of houses in St James Square, London, that it was held 'a building might be of special architectural or historical interest by reason of its setting as one of a group'. While 'group value' might refer to a terrace or a square, consideration was by then also being given to a much broader concept namely the Conservation Area. At this point, we can refer to a further hero of the conservation movement, Duncan Sandys, MP. Despite being a government minister, he had been responsible in 1957 for the creation of the Civic Trust as an umbrella organisation for the growing number of local civic societies. It was the Civic Trust that in January 1967 published *Preserving the Architectural and Historical Scene*, the first full study of the Conservation Area concept, supporting the protection and even enhancement of whole areas of architectural or historic interest. It was then Duncan Sandys who, as a backbench MP, introduced a private member's bill and gained government support for this so that it became an all-party measure, receiving Royal Assent in July 1967 as the Civic Amenities Act.

Two prejudices were overcome at one time with this Act. First came the recognition of the historic and aesthetic value of whole areas of buildings and the spaces which they enclosed or which provided their setting. While the individual buildings might not be of adequate merit for listing, together they had townscape value and were worthy of retention and protection. The second prejudice involved who should be responsible for the designation. Listed building approval required the decision of a government minister at national level. By contrast, Local Authorities were deemed capable of designating Conservation Areas. The initial goals defined were remarkably vague. The task given to Local Authorities was 'to determine which areas of special character or appearance it is desirable to preserve' (Civic Amenities Act 1967, Section 1). The words 'special', 'character' and 'appearance' were not clarified. Nevertheless, from an initial four areas designated in 1967, the concept has proven exceptionally popular and today there are over 9,000 Conservation Areas.

Effectiveness of the Legislation

Over time, the machinery for the protection of the built environment has developed from its initial passive reliance on voluntary acts by private owners to the imposition of rigorous statutory controls and sanctions. However, much still depends on the will to use these. Currently, 94% of applications for listed building consent in the UK are granted. While the rate of destruction of listed buildings slowed down dramatically, from an average of 500-600 applications for demolition a year in the 1980s to c. 200 today, English Heritage estimated in 2000 that, 'nearly 4% of Grade I and Grade II* listed buildings and structural scheduled monuments are at risk through neglect and decay ... a similar proportion of Grade II buildings are at risk from neglect' (English Heritage, 2000:13).

A key reason for the reduction in the number of applications to demolish has been the fall in the number of applications from Local Authorities themselves, from 1 in 3 for 1977 to only 1 in 33 for 1994. However, there continues to be only limited monitoring of whether the advice given on listed building consent applications by English Heritage and the equivalent bodies in the other parts of the UK is followed, and little information is available to judge Local Authority enforcement of listed building controls or monitoring of change in Conservation Areas. In recent years, there has been a substantial decline in the number of specialist Local Authority staff working as conservation officers in identifying, evaluating and monitoring relevant buildings and areas, and supporting owners: 'even when qualified conservation officers are employed, they may be in junior positions that give them little or no opportunity to advise on decisions critical to the historic environment' (English Heritage, 2000:34). Equally, the legislation is not as effective as it might be. Too often conservation area designation achieves little (English Heritage, 2000:35). This is reflected graphically in the current UK epidemic of do-it-yourself activities, such as the installation of uPVC windows and doors, which can be a major problem in conservation areas.

Ultimately a building must remain economically viable. Frequently this involves seeking a new use and accepting changes to the fabric to make this possible. Conservation policy has never sought to prevent change, rather to slow it down and to bring it under more control. Alongside statutory controls is the need to develop and retain the skills and means to manage conservation and change. These vary from the entrepreneurial role in economic regeneration to the specialised skills required to deal with necessary works of sensitive repair, replacement and adaptation. Given the key role of the architect in the latter, it is horrifying to realise that, after 1945, British architectural schools abandoned the teaching of architectural history and traditional building techniques, concentrating instead on modern design and methods. The training of new conservation architects and the re-discovery of traditional craft skills has been a slow process in Britain since the 1960s but conservation still does not form an integral part of professional training.

Alongside this issue lies the need to educate private owners on the advantages of traditional building methods and materials, supported by advice and grant aid, with statutory powers kept as a last resort. The weakest point in the machinery for building conservation in Britain remains the tax system. In Britain, almost alone in the European Union, 17.5% VAT is imposed on the repair of listed buildings and those in conservation areas, while demolition, new buildings and alterations are zero-rated.

Conservation Areas also include numerous town centres and other areas which have come to be known as 'historic urban quarters'. Frequently, these have been economically run down and statutory responsibilities have been largely irrelevant. Instead, the public sector has had to take a much more active role in encouraging their regeneration, but without the financial resources to carry out the necessary maintenance and improvements. Public bodies have had to learn how to work in partnership with the private sector and to create the machinery which makes this possible. They have also had to convince the private sector that conservation would still permit a proper financial return.

In 1968, the government commissioned studies of four historic towns: Bath, Chester, Chichester and York. The intention behind these was not just to focus on the four towns concerned, but to provide a level of general guidance that might be applicable in other towns where a combination of run-down old buildings, traffic and pressure for new development posed problems. The long-term effect has been that the public sector has had to continue to fund environmental improvements in the open spaces within the Conservation Areas and also to provide grant aid to private developers to ensure they obtained an acceptable return on their investments.

Retail centres, like the four towns mentioned above which have all developed into major tourist destinations, have provided one type of problem. Ex-industrial quarters have been another matter. Dockland areas have been fortunate in benefiting from an apparent world-wide phenomenon in favour of regeneration through mixed use, although the sheer scale involved ensures continuing problems. Other zones, such as Little Germany in Bradford, the Jewellery Quarter in Birmingham and the Lace Market in Nottingham, have been much more difficult to return to life. The consequences of action to date are discussed in the next section.

The Problems of Success

> I have a recurring nightmare, that sometime in the next century the entire country will be designated under some conservation order or other. The people actually living there will be smothered with bureaucratic instructions limiting their freedom. We will have created a sanitised, bureaucratised and ossified countryside out of something which has always been, and should always be, a product of the interaction of man and his environment as time goes by.
> Nicholas Ridley, Secretary of State for the Environment, speech to National Association of Conservation Graduates, January 1988.

Listing

The fourfold increase in the number of listed buildings since the 1970s means there are now over 520,000 listed buildings in the UK. The figure would be much higher had the counting procedure not been changed and terraces (e.g. the Royal Crescent in Bath) described as one item. Currently, listed buildings represent c. 2.25% of the entire building stock. The limited resources currently available for identifying, surveying and recording additional buildings makes it unlikely that a further full national re-survey will take place. The main threat is no longer major re-development, particularly as many strategic road schemes have been abandoned, but of piecemeal destruction through insensitive conversion and the addition of unsuitable extensions. Nothing lasts forever. Uses change, making adaptation essential if the buildings are to remain viable, while the structures themselves will eventually decay and crumble.

Reality lies in the negotiations over the details between developers and conservation officers and the appreciation by both sides that the building will be at risk if no viable use can be agreed. The end result is invariably compromise - the 'nibbling away' of the original historic fabric. To the commercial imperative has now also been added the provisions of the Disability Discrimination Act 1995 for enhancing access and circulation (Foster and Coles, 1996). Listing has never meant the permanent freezing of a building. The challenge in responding to the bulk of proposals to demolish listed buildings is to seek ways to enable an historic building to meet the needs of modern society whilst retaining its basic integrity.

Conservation Areas

If the long-term problems for listed buildings seem insurmountable, it is the issues raised by the enhancement of Conservation Areas which have caused the greater controversy. Since the first four Conservation Areas were recorded in 1967, the number in the UK has soared to more than 9,000, a threefold increase since the first designations were completed in 1971. In addition, the boundaries of many of the original areas have been extended. This increase has been steady rather than sporadic and is still continuing; most owners view designation as preventing unwanted change and enhancing property values.

A key impact of granting Conservation Area status to existing residential areas, or to the adaptive re-use of old industrial areas for residential purposes, has been the apparently inevitable change in the social character of the area through 'gentrification', or 'snob zoning', where lower income residents and uses are replaced by higher income residents and the facilities that respond to their needs. The physical shell of the area remains but its traditional character is destroyed. The equivalent also appears to happen in historic town centres and, particularly, in other types of historic urban quarters. Their historic, industrial and cultural associations should be viewed as being as important as the physical fabric to their identity and unique qualities.

Frequently, however, these areas have become economically rundown before designation as Conservation Areas. The planning authority actually uses designation to enhance its controls over development. This is a two-way process. On the one hand, designating the area brings a responsibility to maintain it and so involves the public sector in seeking out viable new economic use. On the other hand, the revitalization of these areas is often central to a broader strategy for the economic regeneration of the town or city itself, bringing in workers, shoppers, tourists and residents. All too often the demands of this two-edged sword are met by developing new functions and activities involving heritage tourism and leisure use. It is this process which has led to such heated debate in Britain over the rise of the so-called 'heritage industry'. 'Nineteenth century industrial 'heritage' centres are beginning to look very similar with their newly-laid cobbled streets, catalogue 'heritage' street furniture, retro architecture and chain retail outlets ... historic settlements and urban areas are seen as assets, readily transformed into products that are sold to consumers seeking an 'experience'' (Orbasli, 2000:2). Whether a response to tourist demand or part of a strategy to revive a run-down historic area, it has encouraged the provision of environmental improvements, heritage attractions, on-site interpretation, events and private-sector support facilities including gift-shops and tea-rooms. 'How within a single generation fears *for* the survival of an urban past could become *fear of* the consequences of its conservation ... heritage has become an urban resource, ... which shapes not merely the form but the functioning and purpose of the 'commodified city'' (Ashworth & Tunbridge, 2000: 2).

From the Past to the Future

The National Heritage Act 1983 switched responsibility for many matters relating to conservation from the Department of the Environment to new bodies, best known by their popular names as English Heritage, Historic Scotland, and CADW in Wales. For the first time, these areas were separated off from general planning responsibilities. 'It was widely taken as implying greater emphasis on marketing the historic environment rather than conservation for its own sake' (Ward, 1994:211). This trend of separating policy for heritage from the rest of the planning process continued with the establishment of the Department of National Heritage in 1992 (subsequently the Department of Culture, Media and Sport). The jury is still out on the long-term consequences of this process. The actual relationship between conservation and change seems to be locally determined to a substantial degree. The UK is also beginning to be subject to new government policies aimed at urban renewal, as part of an objective of reducing demand for new residential development on greenfield sites. This will have a major impact on the quality of the built environment. Continuing popular affection for historic townscapes should ensure a positive role for the built heritage in this. Nationally, we are currently witnessing a rush to 'loft' living in converted properties in historic urban quarters, further 'gentrifying' these areas. The issue of heritage preventing essential change, making the modernisation of cities impossible may, to many involved in conservation, sound too much like developers crying wolf. The reality has been and always will be that for conservation to succeed, it must be economically viable.

References

Ashworth, G. J. and Tunbridge, J. E. (2000), *The Tourist-Historic City: retrospect and prospect of managing the historic city*. Pergamon Press, Oxford.
Brown, Baldwin G. (1905), *The Care of Ancient Monuments*, Cambridge University Press, Cambridge.
Department of the Environment and Department of National Heritage (1994), *Planning Policy Guidance: Planning and the Historic Environment* (normally known as PPG 15), HMSO, London.
English Heritage (2000), *Power of Place: the future of the historic environment*, English Heritage, London.
Foster, L. and Coles, A. (1996), *Perspectives on Access*, Museums & Galleries Commission, London.
Harvey, John H. (1993), 'The Origins of Listed Buildings', *Transactions of the Ancient Monuments Society*, vol. 37.
Hunter, M. (ed) (1996), *Preserving the Past: the rise of Heritage in modern Britain*, Alan Sutton Publishing, Stroud.
Orbasli, A. (2000), *Tourists in Historic Towns: urban conservation and heritage management*, E. & F.N. Spon, London.
Ruskin, J. (1880), *The Seven Lamps of Architecture*, 2nd edition. This particular quote also appeared in the 1st edition, 1849. Allen & Unwin, London.
Studdards, R. and Hargreaves, J. (1996), *Listed Buildings*, 3rd edition Sweet & Maxwell, London.
Ward, S. W. (1994), *Planning and Urban Change*, Paul Chapman Publishing, London.

3 The Conservation of the Built Environment in Sweden

BENGT O.H. JOHANSSON

Sweden is a part of the Scandinavian peninsula, which is dominated by forests and mountains, particularly to the north. It is one of the largest countries in Europe, but has one of the smallest populations: only about 8.9 millions of which 80% live in the southernmost part of the country.

Only in the seventeenth century did Sweden occupy an important role in the political power struggle of Europe and only recently did it become (a somewhat hesitant) member of the European Union. In spite of that Swedish culture has always been open to influence from the world's more dominant cultures: Germany and to some extent The Netherlands in the sixteenth and seventeenth century, France in the eighteenth, Germany again in the nineteenth century and since World War II, the USA and the Anglo-American culture.

Construction of a National History

> The island of Atlantis is not something made up by Plato, nor is it America, nor Africa, nor the Canary islands, nor is it sunk in the sea, but it is the same as now is called Sweden. (Olof Rudbeck, 1679)

When the Swedish King Gustavus Adolphus (1595–1632), well known in the history of Europe because of his prominent role in the 30-years war, was crowned in 1617 he chose to appear dressed as the gothic King Beoric, who according to current myths had led the Goths in the conquest of Northern Europe. Gustavus Adolphus would, as we know, soon try the same. A proclamation issued at the coronation stated: 'The right homeland of the ancient Goths is Sweden, the Swedes and the Goths not standing behind anyone when it comes to manhood and loyalty'. In this he was drawing on old pretensions, internationally launched as early as 1453 where at the Council of Basee a Swedish bishop tried to elevate his rank by insisting, quoting among others Jordanes who had positioned the Goths homeland in the far away island of *Scandza*, that he represented a people descended from the ancient Goths who had once conquered Persia, Troy, and England and reigned over Rome and Spain, just to mention a few. No kingdom, he concluded was older, mightier, better or more noble than Sweden (Broberg, 1991:297).

This amazing story was further elaborated by Swedish scholars in the sixteenth and seventeenth century claiming that all the people of the world originated from the south of Sweden (Götaland – note the similarity between Göt and Goth) and that the first Swedish kings descended from Magog, the great son of Noah or that the lost Atlantis was really Sweden from where a few hundred years after the Flood mighty armies conquered the world.

The alleged history played an important role in the seventeenth century when Sweden had emerged as a dominant country of the North. Not only did the government compete with Denmark in the domination of the Baltic region, it actively tried to counteract any pretensions the Danes might have to a more glorious prehistory. Thus there was a need to arrange for the documentation and preservation of ancient monuments that could be reminiscent of the glory of the forefathers. In fact the Danish king had already in 1622 asked for reports on ancient monuments, runic inscriptions and the like that could shed light on Danish history. That was far more ambitious than the inventory of runic inscriptions that Gustavus Adolphus's tutor Johannes Bureus had been commissioned to undertake in 1599 (Schück, 1932).

King Gustavus Adolphus was not only a skilled master of warfare, he also proved to be a sophisticated organiser of civic reforms. He not only militarised Sweden, he founded 14 new towns and he, or rather his Lord Chancellor Axel Oxenstierna, reformed the government itself, the courts, the universities, and higher education. Cadastral maps of Sweden should be drawn in order to get a better basis for taxation. Regional administrative boards were created (which in principle still existed) as well as new central boards or 'colleges', *kollegien* as they were called. The King's Council was thus reorganised with the creation of five colleges with different and well-defined responsibilities each led by a high-ranking official. No wonder then that one of these five modern colleges was indeed the College of Antiquities. This task was to be led by a National Custodian of Antiquities who was charged with organising the nation's past in order 'to refute the Danes who are praising themselves for their monuments'. The King's old teacher Johannes Bureus became the first Custodian in 1630. It so happened that at the same time his first cousin was entrusted with the task of cartographing Sweden in series of cadastral maps.

Figure 3.1 Olaus Rudbeck dissecting geography
to demonstrate to ancient geographers Sweden as the forgotten Atlantis, the right homeland of civilisation
Source: Title page from Rudbeck's 'Atlantica', 1679

In 1666 a regular heritage law followed, possibly the first at least in the Western world after the city of Rome's soon forgotten mediaeval laws on protection of the antique ruins (1363). The Swedish law stated that nobody should be allowed to destroy old buildings, castles, forts, or cairns regardless of the size of their remains. Nor should stones with runic inscriptions be disturbed in any way, including removal from their site of origin. Churches were among the items that should be preserved according to the law.

This interference with the business of the church may be astonishing to other traditions of heritage management, which expressly exclude churches from those kinds of properties that could be protected by heritage laws. We must however remember that the Swedish church was gradually nationalised from 1527, as a part of the Protestant reformation. A new canon law stated in 1571 that the Bishops were responsible for the upkeep of the churches and their movables. In 1776 this power was transferred to the Royal Board of Building (Schück, 1932). This early central control of all churches was to become important in the development of heritage management in the nineteenth century when it was gradually transformed from aesthetic and liturgical control into a control of heritage matters.

It is also interesting to note that already in 1666 there was a clear understanding that a heritage law should include definitions of that heritage and also be followed up by inventories. A few more points of importance for the future should be noted. First the interest in the immaterial part of the heritage and secondly the notion that runic monuments should be left at their original site. This understanding, that there is an important connection between the monument and its site, is a cornerstone in modern heritage management, elaborated in for example the standard setting Venice Charter (1964). Thirdly, that the interest of the nation took over the interest of landowners, no compensation was offered for the preservation order on monuments given by the law. A forth point is that the objects mentioned in the law were automatically protected, as no listing procedure was required. And last, that there is no fixed time limit for how old a monument must be in order to be protected as a heritage. This, for heritage laws of other nations rather unusual omission, has been important in the development of heritage management in Sweden. The past is understood as an *elastic* concept. This was surely unintentional from the start but has allowed a continuous modernisation of heritage management according to the development of the interest in the past.

There is no comprehensive record of how the law worked in practice and only a few examples have been reported. The Government interfered during the seventeenth century in a couple of instances when otherwise prescribed street regulations were threatening mediaeval monuments in the towns of Vadstena and Skänninge in the province of Östergötland. In 1667, the Government decided to compensate the owner for continuous maintenance of a barn in the province of Dalecarlia where the father of the reigning dynasty and founder of the independent national state supposedly had been hiding from Danish soldiers in the sixteenth century. (No obligations for maintenance had been incorporated in the law, just one year old at the time.)

The eighteenth century and the years of Enlightenment did not take much interest in the old tales of the mystic Goths. The official task of defending ancient monuments, and not least collecting historic finds for the National Museum of History, became a one-man task up to the late nineteenth century.

The Awakening of Public Opinion

A revival of Gothicism occurred in the early nineteenth century in the wake of the nationalism following Napoleonic war around Europe. Again, the glorious past was evoked - this time however in a peaceful way. The cultural unity of the Nordic countries was stressed, the Vikings were a common heritage, and the Nordic peoples should from now on become brothers. The heathen Gods were idealised and depicted like Attic heroes; their allegedly strong morals and simple manners were praised and held up as models in education. Centre of this movement was 'The Gothic League' established in Stockholm in 1811 with a cause: to revive 'the old Goth spirit of freedom, manly courage and integrity' (see for example Ohlsson, 1993). Leading members were some of the more talented Swedish poets and artists, but also the National Custodian of Antiquities was one of the League's most enthusiastic spokesmen. One of the initiatives taken was to print a new journal for 'amateurs of the Nordic past' (N.B. 'Nordic' not Swedish).

The national aggressiveness that had accompanied the earlier Gothicism was gone, Finland was lost to Russia in 1809 and it was now time, in the words of the much admired poet and member of the Gothic League Esaias Tegnér, to regain Finland *inside* present borders. That is to cultivate and to reform the country without further expansion. A new popular movement was born, the Scandinavism, and consequently the Danish society for the preservation of ancient monuments and historic documents was named a *Nordic* society and had many Swedish members. Esaias Tegnér, mentioned above, was one of them.

Figure 3.2 In bygone days the Goths used to drink from horns
Students from the University of Uppsala meeting in 1842 among ancient monuments in Old Uppsala, the site where – according to Rudbeck – the founding of the Kingdom of Sweden took place
Source: Rudbeck's 'Atlantica', 1679

The activities of the league and its followers led to a modernisation of the heritage law in 1828 which was now brought up to contemporary standards by expanding the concept of ancient monuments from the oldest and most remarkable, from the memories of the ruling class to more ordinary monuments that could give information on the nation's history at large. It served – or such was the intention – as a reminder of the half-forgotten law of 1666. That is why it was ordered that the law should be read out in all churches to the congregations every second year. But the law also took a step backwards (from the point of view of the preservers) by introducing a principle of compensation to a landowner who was denied permission to remove a monument.

The late romanticism of the mid eighteenth century brought 'the people' definitely into focus. It was the people and their traditions, not the rulers that constituted the nation and held it together. This should be the reason for preserving the heritage not only of Sweden as such but also of the different provinces.

> Our history has grown from being only the history of the state or a history of the public life. Lately it has developed into a history of the particular, of life so to say even inside the house, of peoples' habits and customs, of their culture during different eras.
> Swedish Society for Ancient Monuments (1872)

1856 was the year when the first regional heritage society (or as it was called Ancient Monuments and Popular Language Society) was founded in the province of Närke in central Sweden, soon to be followed by other local associations. The founder was Lord Nils Gustaf Djurklou, his title bearing witness to the fact that the heritage movement was still initiated by the upper classes. In the next ten years another ten regional societies were founded, each of which represented a certain province, as an emerging concept nearly as important in the mind of the societies as the nation itself. These societies were interested in both the tangible and the intangible heritage and (following the ideology that a nation/province consists of one people and one language) wanted not least to record the specific dialect of their district (Antiqvarisk Tidskrift, 1863-1864). Another important initiative was that of Arthur Hazelius who in 1891 opened Skansen in Stockholm as the first open-air museum of folklore and relic vernacular buildings. It was conceived as a mini-Sweden, the Sweden that was vanishing before the modern eye, a mini-country with old-fashioned environments from all the regions of Sweden. Hazelius did not set out to show the monumental Sweden but 'Sweden as it was' before the industrial revolution as the 'uncorrupted' soul of Sweden before modernisation. The initiative was supported by the Royal Family and the higher level of society and became immensely popular not least because Hazelius was exceptionally gifted in blending education and amusement. The name of *Skansen* in Swedish simply means 'the Fortress' (which the site had been) but this name has been reused all through the twentieth century as a synonym for open-air museums all over Europe. Hazelius also took part in the founding of the Swedish Ancient Monument Society in 1869, a society that was asked for at a Scandinavian Artists Meeting the same year, which demonstrates that ancient monuments and traditions were still an important issue with artists at the time.

The development during the latter half of the nineteenth century could be viewed as bringing mounting tensions between the emerging fields of heritage, between centre and periphery, amateurs and professionals, and architects and archaeologists. There was, on the one hand, the Royal Academy and the National Custodian in the capital who was mostly interested in archaeological finds and mediaeval churches, on the other hand there was a growing movement of amateurs from the provinces with broader interests in heritage. There were on the one hand the beaux-arts trained architects at the Royal Board of Buildings in the capital, on the other hand self-trained experts in mediaeval architecture and culture. Alternatively, on the one hand there were the architects with their natural interest in preserving buildings, on the other hand the National Custodian with his still more limited interest in archaeology and ruins.

The regional heritage societies soon started to develop provincial museums of archaeology and folklore. They fought a bitter fight with the Royal Academy and its secretary the National Custodian of Antiquities over the rights to archaeological finds which they believed were of better use in their region of origin than in a museum in Stockholm where they could not as well foster 'a spirit of nationality' among the population. From the 1890s onward the heritage movement spread fairly quickly over the country. Local heritage societies were (and are) formed in each and every corner, many of them with museums or 'mini-Skansens' of their own. In 1922 there were c. 150 societies (SOU 1922, pp. 222-372), in 1940, 715 societies counting c. 125,000 members which societies today (1995) have expanded into 1,600 societies with c. 400,000 members (5% of the total population of Sweden), owning close to 3,000 heritage buildings.

(Re-)Establishing the National Framework

Using a rather free interpretation of the wordings of the Ancient Monuments Act of 1828, the lonely National Custodian of Antiquities managed to create a network of representatives, or ombudsmen, in different parts of the country; these were given the power to report, document and intervene when ancient monuments were threatened or ancient finds were unearthed. These ombudsmen would typically be drawn from the leading members of the regional ancient monuments societies. It would have been these societies that influenced the members of parliament to vote for a new legislation for ancient monuments, much stricter and at the same time broader in scope. The new law that emerged in 1867 (two years before the constitution of the Ancient Monuments Society) stated that *all* ancient monuments were protected by the law (regardless of importance), and that no compensation was to be given to landowners for the restrictions in land use that derived from the law (that is: back again to the principles of the seventeenth century which were recalled in the preliminaries). It followed that registration of monuments was not a prerequisite for protection. Instead the law enumerated which types of monuments should be included, a list that in principle prevails today. The Act has been changed and expanded several times since but its main principles still exist, the latest amendment being a recent inclusion of old place-names.

The National Custodians of the eighteenth and nineteenth century had mostly acted as head of the Central Museum of Antiquities and thus were mainly absorbed in enlarging their collections. During the industrialisation of Sweden during the latter half of the nineteenth century they were however more and more involved in what was called 'the outward heritage management'. At the beginning of this century it was obvious that the Swedish heritage administration must be thoroughly modernised. Following a special committee's advice (SOU, 1922), a central administration under the National Custodian was formed with several departments including the Central Museum of Antiquities. The system of regional ombudsmen mentioned above was formalised into regular County Custodians partly financed by the Government, partly by the Counties. The County Custodians were at the same time head of a regional museum and secretary to the local heritage society just like the National Custodian who in his turn was head of the Central Museum and at the same time secretary to the Royal Academy.

This tradition of intermingling heritage management and popular education and research has been of great importance to the rooting of heritage management in society as well as for a sharing of power between the central state and the regions.

Today the responsibilities for cultural heritage administration in Sweden have become even more decentralised. Enforcement of the specific legislation related to historical monuments is since 1977 (with the exception of state owned protected buildings) delegated to the County Administrative Boards, which have a professional staff of cultural heritage officers in their departments of environment or similar departments. These County Administrative Boards play an important role in administering the cultural environment through instruments other than heritage laws, including planning and economic incentives. Documentation and information issues are handled by county museums, one of which is found in every county. These museums receive as before state grants but are governed by municipal authorities and local heritage societies.

In the planning system, Swedish municipalities have the over-all responsibility for the conservation of the cultural environment. Some of them, especially the larger ones such as the municipalities of Stockholm,

Göteborg, Malmö and Norrköping, have city museums with specialised conservation officers. Smaller municipalities co-operate with the regional museum. A few of them have employed a heritage officer of their own who is usually attached to the Municipal Board of Culture. As a general principle The Act of Planning and Building prescribes that buildings and areas of cultural values shall be preserved. The cultural heritage can be protected by detailed regulations in local plans.

Ancient Monuments

Current heritage legislation shrewdly defines ancient monuments in a rather circular way as abandoned monuments from 'ancient times'. It can also be a place or an immovable natural object with ancient traditions or tales connected to it, for example a large stone, a spring or a peculiar tree. There is no time limit for the term 'ancient', ancient being what people commonly think of as ancient. An ancient monument is not only a 'monument', it can also be something that is the result of an activity, e.g. remains of a settlement in the form of cultural layers. A most important reform is the ruling since 1942 that an ancient monument belongs to an environment, which is regarded as an integrated part of the monument. The extension of the environment accommodates the need to understand the individual monument and to protect it. One of the main purposes of the Act is to regulate the landowner's or any entrepreneur's rights and obligations regarding ancient monuments. Such a monument or site may not be taken away, covered or otherwise damaged or changed without permission from the competent authorities. Permission may be granted if the importance of the monument is not proportionate to the importance of the suggested enterprise. The permission may be given on the condition that the enterprise pays for scientific documentation before the disturbance of the monument. It is also stated that the entrepreneur should contact the authorities in order to discuss possible measures to avoid conflict with the monument or site. If a previously unknown ancient monument or site is discovered during exploitation, the work on the site must stop and permission sought for continuation. In a recent amendment to the law the requirement is added that any large-scale exploitation of land must be preceded by special archaeological surveys at the expense of the enterprise. This can be described as a specific part of an Environmental Impact Assessment.

Following this law, the National Heritage Board, which keeps the central register of ancient monuments and sites, has conducted systematic surveys of ancient monuments and sites all over the country since the 1940s. Results are also recorded on the detailed economic map of Sweden, a map that is supposed to be well known to landowners and entrepreneurs.

Historic Buildings

Contrary to the protection of ancient monuments, the legal protection of buildings is by tradition weak in Sweden. An exception to that rule is that all churches belonging to the Swedish Church are protected together with their movables. The law has been effective since 1759 and was usually meant to guarantee that the churches be architecturally worthy of divine services. At the end of the eighteenth century this law began to be used also to protect cultural heritage, which is nowadays the only reason for keeping state control. As the Swedish Church gradually is liberating itself from the State the law has been relaxed. Today only churches built before 1940 are automatically protected. More modern church buildings have to be individually listed by the National Heritage Board. Since the Swedish Church was totally separated from the Government by the year 2000 it may become an issue as to what extent the protection can be upheld. So far the Church itself has been able to pay for restorations partly due to the fact that it had the right to tax people belonging to the congregation (most Swedes were members). This right has disappeared with the separation. The matter will depend on the Government's willingness to pay compensation in the future.

In 1920 the Government ruled that culturally valuable buildings owned by the state should be treated with respect for the cultural values. The Central Board of Antiquities became responsible for suggesting to the Government which state owned buildings should be listed. With regard to private and municipal buildings, these could not be listed before 1942 and only with the owner's consent. It was felt that listing otherwise would be rather too large a reduction of the owner's right to dispose of his property. Only in 1960 did the law open up for listing against an owner's wish on the condition that the Government compensates the owner economically. The law also states that listing can only concern buildings of outstanding cultural importance or belonging to an assembly of such importance. Thus there are no different grades of listing in Sweden. There is also no defined age eligibility for a building to be listed. When a building is listed the County Administrative Board has to define in what respects the building is protected (for example façades, special interiors or details) which also calls for a thorough documentation prior to the decision. The site of the building may be included in the listing. The law also covers parks and constructions like bridges but movables cannot as yet be protected even if they are historically connected to the building.

The Swedish listing process is, as can be understood, a rather complicated matter, apart from the fact that the very restricted amount available for economic compensation makes it even more difficult. Tax reduction although frequently proposed in discussions has proven to be a political non-issue much to the regret of conservators. (The main argument against, which all political parties represented in Parliament hold true, is that tax reductions of such types would complicate the system and create new loop-holes. Another argument voiced from the political left is that special tax reductions favour those with higher incomes.) Thus there are only about 2,000 listed buildings in Sweden. Just a few of them (c. 50) were built after 1920. Listing of areas is also extremely rare, as each building has to be listed individually.

According to the law on nature conservation The County Administrative Board can issue preservation orders on areas of historic interest in the countryside. In circumstances when preservation orders have been issued, the area has however been of dominant interest to the conservation of nature. Obviously this has to do with the fact that cultural and natural environments are treated in different boards under different ministries. The law has, however, recently (June 1998) been revised and turned into a general law on environmental protection and amended with a special paragraph on the protection of cultural heritage areas, which is meant to improve the situation.

Planning and Planning Laws

Planning is a monopoly of the municipality and the national government can only interfere in certain cases. These cases are defined in the Act on the Management of National Resources. According to this Act, national interest must be respected. Those interests are mainly of three kinds: first, matters of public health; secondly, interests of exploitation on an national scale like building of railways, national roads, mining; and thirdly, interests of preservation like cultural and natural resources and the need for recreational spaces, especially in larger conurbations.

Regarding cultural and natural heritage, the law states that areas of national interest must be protected from enterprises that may substantially damage the values that constitute that very national interest. It is up to the National Heritage Board and The National Agency of Nature Conservation, respectively, to point out these areas. On the list drawn up by the Heritage Board there are about 1,700 areas. These are selected in order to reflect different aspects of Swedish history, as the Heritage Board in a dialogue with regional boards and museums conceives it, and they cover development since the Stone Age up until the city renewal projects and satellite towns of the post-war area. In the list arguments are given for the selection, which are clearly dominated by a materialistic view of history. It could be argued that the list is conventional with a tendency to stress environments left behind by development, environments that today appear idyllic whatever grim memories they may harbour. It is about homogeneity rather than disorder. Whose history is thus left behind? What ugly skeletons are hidden? Is the list still upholding a harmonious view of Sweden as a successful national state? It is however interesting to note that scholars, local heritage societies or indeed the municipalities have not yet really challenged this list, which is now close to 15 years old. In fact professional suggestions to 'beautify' a more ugly or trivial modern heritage like housing projects of the 1960s or industrial zones from this century without heroic buildings, usually fail to raise an understanding by the general public.

According to the Act on the Management of National Resources the municipalities must show in a comprehensive plan how they plan to treat the areas of different national interest. Those plans are scrutinised by the County Administrative Board, which when in doubt, should consult the National Boards concerned. About 50% of the municipalities have produced cultural heritage programmes wherein local cultural values are described. These programmes are used as non-binding guidelines for the planning process and as educational tools for the public.

As was stated above and as a general principle, The Act on Planning and Building prescribes that buildings and areas of cultural values shall be preserved. Planning permission for the removal of a culturally valuable building may be withheld. The cultural heritage can be protected by detailed regulations in local plans. As in the case of listing as described above, these regulations may not however substantially diminish the value of the property without the owner's consent. If the municipality nonetheless wishes to proceed, full compensation must be paid. There are no known cases when this has been done.

The Possible Problems of Success

While the United Kingdom and The Netherlands may have problems with a heritage sector that may be perceived to be too dominant this is somewhat harder to argue when it comes to Sweden. True, there is a certain dissatisfaction among younger, internationally oriented architects, especially so among those who are attached to architectural schools. These circles propose more daring modern architecture for in-fill situations in existing townscapes and in the renovation and reconstruction of old buildings. The same people could however also be heard to argue for more respect for the internationally oriented architecture of the 1960s, the period which is still very much disliked by contemporary politicians and the general public. This could partly be dismissed as a matter of taste being more about what type of environment to conserve than about real distress with too much conservation.

Three more substantial questions on the consequences of the spread of the conservation movement could however be raised, first, about the reuse of conserved buildings, secondly about conservation/preservation as a burden on public economy, and thirdly about conservation/preservation as a restriction to development.

As everywhere else there are of course problems of finding new uses for heritage buildings situated in areas with economic problems. Many smaller municipalities are at the moment tearing down housing projects from the expansive 1960s. The industrial heritage, so much appreciated these days, is particularly difficult in this respect with many industrial buildings being hard to rehabilitate for new uses.

As may be understood by the modest list of ten industrial heritage complexes asked for by the Government, Public Authorities are very hesitant to take on any economic burden to preserve the heritage. This was confirmed by a recent inquiry from the Association of Swedish Municipalities from which it was clear that the costs of preservation is a main reason for municipalities not engaging themselves actively in the issue.

The third question is maybe still harder to answer with any precision when it comes to what degree development is restricted by preservation. No serious complaints from real estate managers or municipalities have really been heard. It is true that preservation of the old town of Stockholm is radically restricting the opportunities to build new buildings. Still the Old Town is economically vibrant. Development thus can take many forms and a real estate developer may have many options about where to invest. If retail centres and large parking areas are no longer allowed in historic town centres there is always an option to build such centres down by the highway.

The building stock in Sweden is young, especially in comparison with the UK. The bulk of buildings come from the period 1950-1975 (close to 80%). Older buildings have as a rule been thoroughly modernised. According to the statistics for living standards in housing, there is practically no accommodation without modern conveniences left in Sweden. The same could be said for offices. The urban renewal of the 1960s hit Sweden particularly hard. Close to 50% of urban housing built before 1900 were destroyed in the process (Johansson, 1997).

The losses, physical and emotional, to the traditional environment that occurred in many European countries as a result of the Second World War led to large post-war reconstruction programmes. This did not occur in Sweden but large-scale urban renewal rolled over the country anyway. The reaction to this in the media and among voters, in the end, became so strong that many municipalities changed political control in the upcoming elections. But the weak legislation for the built heritage and the strong position of landowners in planning prevented undue stress being put on preservation. At the same time it should be confessed that sometimes the widespread popularity of conservation has led to restrictions in the appearance of new buildings that have resulted in unfortunate architectural compromises. A typical example is the development of the 1980s in the periphery of the historic centre of Kalmar, an otherwise unusually well preserved seventeenth century fortress town. Here, a large office complex has been dressed up in 'historic' forms, accepted, one would guess, as a matter of 'post-modernism'.

References

Broberg, G. (ed), Gyllene äpplen (1991), *Svensk Idéhistorisk Läsebok*, Stockholm, p. 297.
Johansson, Bengt O.H. (1997), *Den Stora Sstadsomvandlingen. Erfarenheter från ett Kulturmord* (The Great Restructuring of Cities), Stockholm.
Ohlsson, Per T, Gudarnas ö. (1993), *Om det Extremt Ssvenska* (The Island of the Gods. About the Extreme Swedish), Stockholm.
Rudbeck, Olaus (Olof) (1679), *Atlantican* (1937-1950).
Schück, Henrik, Kungl. (1932) *Vitterhets Historie och Antikvitetsakademien: dess Förhistoria och Historia*, del 1, Stockholm (The Royal Academy, its Prehistory and History, part 1) Antiqvarisk Tidskrift för Sverige, 1-2, 1863 and 1864.
SOU (1922), *Betänkande med förslag till lag Angående Kulturminnesvård samt Organisationen av Kulturminnes-vården*, del 2 (Report regarding Heritage Laws and Management, part 2).

4 The Conservation of the Built Environment in The Netherlands

G.J. ASHWORTH

The Netherlands is a physically small, densely peopled, and governmentally centralised country, with a high degree of social homogeneity, centrally located between the major European cultural blocs of Germany, France and Britain. Since the 1960s it has combined economic prosperity with a strong tradition of social equality and a popular acceptance of quite extensive planning intervention in the public interest. The country has played a major role in European historical events at least since the sixteenth century and has for about 400 years participated in the major European cultural trends, movements and fashions. It could be expected therefore that it would have developed an effective, nationally sponsored system for the conservation of the built environment at about the same time and in a similar way to its North West European neighbours. However its relatively small population (c. 15 million) and cultural vulnerability to outside influence might also be expected to have created distinctive issues.

The Struggle

Although there has been planning intervention in such spheres as water control for many centuries the main barrier to government action for urban conservation in the public interest was the long dominance of the Liberal State. The idea that government action should be strictly limited and not threaten the inviolability of private property delayed effective action in The Netherlands as it did in other European countries. It was a liberal Prime Minister of The Netherlands who declared in 1862 that, 'art is not the business of government: the government is not the arbiter in matters of art or science' (quoted in Ashworth, 1991). Nevertheless, enthusiastic, influential individuals similar to, and usually in contact with, the other European prophets of preservation also promoted a concern for a rapidly disappearing past in The Netherlands in the second half of the nineteenth century. In 1875 Victor de Stuers, a senior civil servant, wrote a hard hitting attack on the small-minded attitude of the government and the governing elite in general who were passively observing the destruction of the past. At the same time the architect P.J.H. Cuypers became the Viollet-le-Duc or Gilbert Scott of The Netherlands with his enormous influence on medieval church restoration (Denslaken, 1994). Thus the long standing and bitter controversy between the 'restorers' typified by Viollet-le-Duc (1875) and the 'preservers as found' represented by John Ruskin (1848), was also present and led to a drastic reconstruction of almost all major medieval buildings to accord with the nineteenth century view of how they ought to appear.

The pioneering role of influential organisations such as especially the *Koninklijk Instituut voor Wetenschappen, Letteren en Schone Kunsten*, (Royal Dutch Institute for Science, Letters and Fine Arts) 1808 was evident here as in the rest of Europe. Similarly private organisations, often under royal patronage, such as the *Koninklijke Nederlandse Oudheidkundige Bond* (Royal Dutch Antiquities Society) 1899 and the *Heemschut* (Society for the Protection of Cultural Monuments in The Netherlands) were also influential. By the end of the nineteenth century, the prophets of preserving the past had convinced at least a substantial and vocal minority that governments should act. The national government in turn, coping with a newly enfranchised population and needing to define a new idea of 'nation' saw advantage in discovering and propagating a newly defined 'national heritage'.

The Dutch, as other European governments, were however slow to make these concerns manifest in formal legislation. In 1875 a *Rijksdienst voor Monumentenzorg* (National Service for the Care of Monuments) was established to make an inventory of the most important national architectural monuments. This was conceived as a non-controversial, cheap and quick operation. A few government employees would for a few years list all preservable buildings in The Netherlands. It was however ultimately neither cheap nor quick and recently also controversial. It was the beginning of a process of government involvement that grew within a century to an enormous and still incomplete commitment far beyond the most optimistic visions of the preservation pioneers.

However, there remained a long gap between the drawing up of lists and the designation of effective protection through legal action. The 1814 Act, inspired by Napoleonic precedents, was unenforceable. A number of Acts in the first fifty years of the twentieth century extended only a partial protective cover and minimal public economic support to the most spectacular monuments. However, the designation 'state monument' (*Rijksmonument*) was created, thus accepting a public responsibility for recognising, designating, legally protecting and ultimately maintaining, at least the most important structures in the built environment. Pressures from both the *Rijksdienst* and from private pressure groups, steadily expanded the definition of monument, lengthened the lists and increased the public funds available for the restoration and renovation of property in both public and private ownership. This was reflected in the establishment of organisations for the preservation of bell towers (1919), windmills (1923), military architecture (1932), castles (1945), and more recently industrial buildings.

The Victory

It became apparent, throughout at least Western Europe, by the 1960s that a new and broader approach was needed. In part this was an aspect of a post-war social-democratic consensus which believed strongly in the need and efficacy of government intervention for collective goals expressed instrumentally in public planning systems. Post-war reconstruction, rapid population growth and a new economic prosperity combined to produce an enormous expansion in demands upon the built environment and especially upon land-use in the cities. This could have resulted, as it did all too often in the United Kingdom, Belgium and France, in the wholesale destruction in the name of redevelopment of the urban built environment and the sacrifice of the city centres to the insatiable demands of the motor car. That this did not occur, with only a few unfortunate exceptions, is perhaps the most important fortuitous and felicitous event of post-war Dutch history. This can be attributed to the existence of a large, effective public planning system, armed with powerful legal instruments of development control. This was supported by a broad popular consensus that not only accepted but even required public intervention to manage the built environment in a public interest that accorded social justice preference over private profit. Thus the preservation of elements within the built environment was combined with concern for the form and functioning of cities.

This thinking resulted in legislation across Europe such as France (1962), Britain (1967), Italy (1970) and even Turkey (1973) (Ashworth & Howard, 2000) all with similar provisions, and justifications. In The Netherlands the *Monumentenwet* (Monument Act) 1961 consolidated previous legislation about individual monuments, increased the possibilities of governmental subsidy for restoration but also recognised the importance of ensembles by creating the *beschermde stads(dorps) gezicht* (Protected urban (village) scene). Unlike some other countries, however, The Netherlands had both the machinery and the political will to implement such policies through the tripartite division of public planning functions between national, provincial and district authorities. Even more important was that monument preservation had become an integral part of urban conservation and concern for the historic built environment had become an element in a much wider approach to cities. Here lies the explanation, not only for the ever lengthening list of monuments and the area conservation of almost all the central areas of almost all Dutch cities but also that no urban motorways and no out of town shopping centres were ever built in The Netherlands.

The consequences of the Act became evident over the next 25 years. There was a steady increase in the number of national monuments designated, in the number of local authorities instituting their own monument protection policies and designating a supplementary category of local monuments (*Gemeente monumenten*) and 'scene determining buildings' (*beeldbepalende panden*), and in the number of towns and villages creating conservation areas as allowed by the Act. Finance for restoration and maintenance was through a combination of relatively generous direct public subsidy and tax concessions. In addition a state sponsored service monitoring and reporting upon the physical condition of monuments (*Monumentenwacht*) was established and is available to both public and private owners of state monuments. A number of local authorities funded a revolving fund for the purchase, renovation and resale of historic property (*Stadsherstel*). Extensive conservation projects such as *Bergkwartier*, Deventer or *Stokstraat*, Maastricht were the showpiece restorations of the 1970s but they were paralleled by less comprehensive programmes in almost every city.

A revised Monument Act in 1988 brought few fundamental changes to this pattern with the exception of decentralising much of the national responsibility to the Provinces and Local Authorities. Furthermore lists were doubled as a result of a national inventory of 'Young Monuments', that is those less than 100 years old which included many 'arts nouveau' (*Jugendstijl*) and interwar (so-called *Amsterdamse* and *Delftse School*) buildings. Not only were there more monuments but significantly more industrial and public service buildings (Figure 4.1).

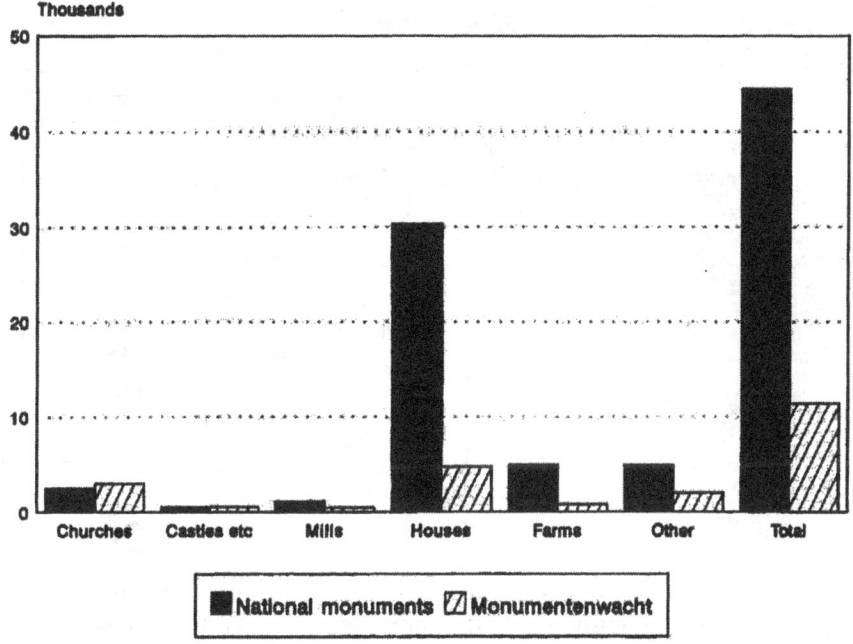

Figure 4.1 Types of monuments and numbers inspected
Source: Statistics Nationale Contactcommisee Monumentenbescherming (NCM) (1999)
    Figure drawn by G.J. Ashworth

The Problems of Success

Forty years of national government financed and encouraged conservation of the built environment and a growing public consciousness of its importance has led to a proliferation of monument designation, of various categories and the extension of conservation areas to cover the central parts of most Dutch towns (Statistics available annually from *Nationale Contact Commissie Monumentenbescherming*). In these respects The Netherlands is comparable to other Western European countries. There is of course still a threat to designated or potential monuments, which are still demolished as a result of the enormous pressures upon land for residential, industrial, commercial and transport uses. There is also a considerable variation in the spatial impact of urban conservation. Some towns, provinces and regions are still much more 'conservation conscious' and thus monument endowed than others (Figures 4.2 and 4.3) as the conservation effort has diffused outwards from the central provinces to the national peripheries. Even more evident however are the many problems that success itself has created which are now the issues around which the conservation debate ensues.

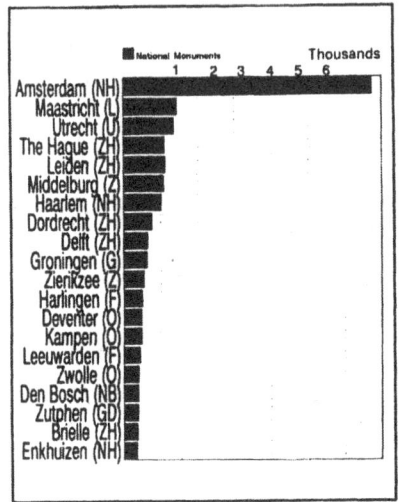 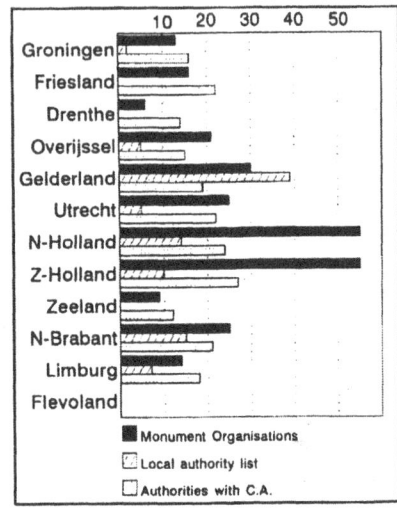

Figures 4.2 and 4.3 Numbers of monuments in towns and provinces
Source: Statistics Nationale Contactcommisee Monumentenbescherming (NCM) (1999)
Figures drawn by G.J. Ashworth

Finance

Amongst the most important of these problems is financial. The simple act of designation and protection from harm incurs few direct costs but it implies a responsibility for maintenance that is both intrinsically expensive, increases steadily as the building ages and unending. The rescue of a few spectacular buildings and their subsequent maintenance was a small item in national budgets but the expansion of the lists and the designation of conservation areas, covering extensive numbers of buildings of little intrinsic merit and often poor physical condition, has steadily increased the burden upon public finances. Not only is the financial commitment to maintain the present stock of conserved buildings a growing problem, there also exists what has been called 'the preservation time-bomb' (Baer, 1995). This occurs because the quantity of additions to the building stock has varied enormously over time. In The Netherlands after several centuries of little building activity, there was considerable expansion in the built environment in the period 1870-1910, stagnation and even property destruction in the ensuing period of wars and inter-war depression, followed by enormous increases in the three decades after 1945. This means that the proportion of the building stock that is eligible, by its age alone, to be considered for preservation will increase approximately five-fold in the next 50 years. Thus if present selection criteria are maintained the number of monuments will increase by the same multiplier with obvious and probably unsupportable financial consequences.

If public resources cannot sustain the growing financial burden, then private resources must increasingly be sought. Private interests and individuals can only be involved in the costs of restoration and maintenance if they obtain a specific benefit, whether economic or psychic, from the preservation. In other words, economically profitable users and uses have to be found for conserved buildings and areas.

Reuse

The expansion of the lists of monuments and the designation of ever-wider conservation areas has not only increased the number of buildings but also altered their nature. Previously monument lists were dominated by large and architecturally or historically significant structures but increasingly the dominant percentage of preserved buildings are small and of only modest architectural or historical interest. Finding a new, appropriate and economically profitable use for a unique medieval cathedral or a seventeenth century palace is not particularly difficult and, even if it were, then vacancy would be a tolerable alternative. Finding new uses for tens of thousands of small premises when these are expensive to maintain, legally or technically difficult to adapt physically to modern requirements and generally located in conserved districts which are inaccessible to motorised vehicles, is a major problem.

The number of commercial and public functions that welcome the benefits of historicity, its conveying of historical atmosphere, social status or patronage or just tolerate the costs, is far less than the number of buildings available. Some types of shop (e.g. antique, antiquarian, arts and craft and the like), private services (especially those professional counter or treatment services requiring customer confidence and credibility) and the 'front-offices' of governments or companies (especially where historicity contributes a style of continuity and status) will occupy, and pay for, conserved buildings or buildings in conserved areas. Tourism needs the heritage product, including the heritage building and district: it is however highly selective in its use. Only the most spectacular examples will be commodified for tourist consumption and tourists only rarely will use old buildings for accommodation.

In practice only two options remain, namely large-scale residential use of conserved premises or extensive vacancy. The main problem with the first is simply numerical. The expansion of the numbers of small, originally domestic, properties has far outstripped the small group in the population prepared to live in, and finance, historic houses. Throughout Western Europe, and especially in The Netherlands, there is a large popular support for the preservation of old houses but an almost equally large rejection of the option of living in them (Rijksdienst Monumentenzorg, 1990). This is partly a preference for the lower costs, convenience, marketability, mortgageability and modern facilities of new housing and partly a spatial preference for suburban locations and their accompanying lifestyles as opposed to the perceived difficulties of living in the central areas of cities where most conservation has occurred. Such attitudes have been encouraged for various reasons by governments, financial institutions and the property development industry who have long favoured 'green' over 'brown' site development. In practice two quite distinct social groups occupy conserved central city premises. First, there are those who deliberately choose historic premises because they acquire benefits from the historicity and are therefore prepared to contribute to the costs. Secondly, there are those who live in such buildings or areas because they are centrally located or just available and who are indifferent to, or even unaware of, the historicity. The second group outnumbers the first in most European cities. However on the occasions when conserved areas prove attractive to the first group, the result is 'gentrification' and social change which may not be considered desirable.

The remaining option, if all the uses in review above fail to occupy the conserved space, is vacancy or underuse, especially of less accessible upper storeys. This leads to the fossilisation or 'museumification' of large areas of the city. It can be argued that what needs now to be preserved in cities is the capacity to change without which cities cannot develop or adapt and thus have ceased, in an important respect, to be cities.

Social and Political Implications

The expanding number of preserved buildings and areas and the increasing role of conservation within local planning has shifted the focus from architectural preservation to the place of fostered historicity within much wider issues. In The Netherlands these have included sharp debates upon the social consequences of area designation and the role of built environment conservation within the functioning of inner cities in particular (see Ashworth & Ennen, 1998). In terms of political legitimation, the origins of the Dutch preservation movement, as in most western European countries, lay in the discovery and propagation of a dominant national ideology designed to encourage the identification of citizens with the prevailing ethos of the state. A liberal, mercantile, protestant, bourgeois, orangist ascendancy is evident in the stress upon the seventeenth century (the concept of 'the golden century') and upon the warehouses, modest residences and trading cities of this period. The difficulty with this 'Vermeer image' of the Holland of a particular period is, as with all heritage, its selectivity. There are now calls for a wider variety of heritages, including the fifteenth century heritage of the Catholic Burgundian southern provinces, and Hansa oriented Northern and Eastern provinces (Ashworth, 1984), as well as industrial and working class heritages, and ultimately also the contributions of more recent immigrant groups, to be included.

From Building Preservation to Heritage Planning

Thus The Netherlands, like much of Western Europe, has moved successively from the preservation of buildings, through the conservation of areas, to the management of heritage. This transformation has taken rather more than a century but has had results that are now quite evident on the urban and rural landscape. A movement that began with more or less emergency measures to save buildings from imminent threat is now a routine matter of professional planning practice. The most important threats that are now experienced are those stemming from the successful incorporation of building preservation into more general planning considerations.

The key concept is integration. The management of heritage, including the conserved built environment, has been integrated into urban and regional policy but also into more broadly based cultural and economic policy of governments at various spatial scales. Evidence of this pervasive significance of heritage is the recent combined report of the ministries of culture, planning, transport and agriculture on the future of urban and landscape conservation (Belvedere Report, 1999). However many of the questions have remained the same over the past 100 years even if the answers are likely to be quite different. Heritage is still a matter of choices and management of heritage involves selection. There is the selection of what is to be considered heritage and treated as such and conversely what is not heritage and can therefore by ignored or demolished. The question, 'who decides?' is thus more important than ever. Choice is also needed in answering the question, 'what is the purpose, or more usually purposes for its conservation?' If heritage is multiused and polysemic then different groups may use the same physical structures for different purposes, again increasing the necessity for management. Finally the questions of, 'how many/ how much?' are becoming more important. When the object of attention was only a few spectacular buildings this question was largely irrelevant but to-day it is not only quite valid but essential to ask how many buildings, areas or entire conserved towns does any society actually require and is prepared to pay for, now and in the future. The struggle for the preservation of the past has been won; the struggle now is with the consequences of that victory.

References

Ashworth, G.J. (1984), 'The management of change: conservation policy in Groningen', *Cities* vol. 1, pp. 605-16.
Ashworth, G.J. (1991), *Heritage planning; conservation as the management of change*, Geopers, Groningen.
Ashworth, G.J. and Ennen, E. (1998), 'De woonfunctie van binnensteden', in M. Seip and G.J.Ashworth (eds), *Binnensteden: analyses van gebruik en beheer* Samson, Alphen aan de Rijn.
Ashworth, G.J. and Howard, P. (2000), *European Heritage Planning and Management*, Intellect, Exeter.
Baer, W.C. (1995), 'When old buildings ripen for historic preservation: a predictive approach to planning', *Journal of American Planning Association* vol. 61 1 pp. 82-94.
Denslaken, W. (1994), *Architectural Restoration in Western Europe: controversy and continuity*, A and NP, Amsterdam.
Nationale Contactcommisee Monumentenbescherming (annual) *NCM monumentenjaarboek* Amsterdam.
Rijksdienst voor het Monumentenzorg (1990), *Monumentenzorg en volkshuisvesting* Woningraad Extra 51.
Ruskin, J. (1848), *The Lamp of Beauty: writings on art*, Phaidon, Oxford.
Steurs, V. de (1875), 'Holland op zijn smalst', *De Gids* vol. 37 3 pp. 320-403.
Viollet-le-duc, E. E. (1875), *Histoire de l'habitation humaine, depuis de temps historiques jusque á nos jours*, Paris.

PART 2
CASE STUDIES

Theme 1: Heritage, Identity and Urban Regeneration

INGRID HOLMBERG

The studies in this section share the intention to address 'place identity' as a central concern for varying actors in processes of urban regeneration. 'Regeneration' has often entailed an interest in the cultural and historic dimension of the urban landscape. Replacing urban renewal within urban planning around the 1970s, it embraces an intended rejuvenation and development of existing physical structures. The main role of conservation in such processes seems to be to define which are the aspects of place identity that apply for historical values and to secure their maintenance throughout the project. In this way, defined historical values of the built physical structures – as parts of the place identity – can become 'constants' to which other actors and other kinds of interventions have to relate.

However, following Kuipers and Ashworth, heritage is 'a conscious use of past associations and references' involving any actor that takes up with the past. The studies encircle and discuss an emerging concern for definitions and enhancements of a historical identity of place, among actors in many different positions: beside local and national heritage authorities, also planning departments and city councils, as well as 'public opinion' as demonstrated in consultations, through amenity societies or urban protest groups. In this way, the workings of urban conservation are set in a broad social and cultural context, and put in perspective of current local planning regulations and forms for provision of heritage financing.

The contributions of this section, consider the theme from two main perspectives. First, there arise cultural questions of how such place identity is conceived, constructed, acted upon and displaced. To this perspective belongs a consideration of the relationship of these conceptualisations of place, to the physical fabric itself. Second, the theme raises questions regarding the effect as well as the very limits of effectiveness of the regulating structure in which conservation and heritage planning is performed.

The first concern is here framed and discussed in several ways. Kuipers and Ashworth expound upon how identity of place can be conceived from a communicative perspective. Stating that place identity seems to be 'itself un-clear, ambiguous, and continually changing', they discern the identification of an 'order of meaning', as one major question for physical planning. In the case study presented, one such order can be discerned ('traditional'), but the case also illustrates how the same physical properties are the object for a conflicting conception ('nostalgic'). The order of such place identity is at the core also in Holmberg's contribution. Focussing explicitly on the shifting conceptualisation of the area of Haga in Göteborg – over time as well as between actors – the aim here is to explore the way the indicated historical values 'correspond to the physical fabric itself, embracing spatial units in certain particular ways'. The contribution of Black likewise illustrates a varying emphasis in the conception of the Lace Market's identity. In this way it opens up a discussion of the effects of the ordering of the place according to certain moral-aesthetic principles of, for example, 'Victorian' or 'industrial', comparable with what is observed as a purified 'workers' identity of place in Holmberg's contribution.

The second concern of the authors is the contextualisation of heritage responsibilities and to consider the structural constraints specific for each case. Black's study contributes here with a detailed discussion of the local regeneration effects for the Lace Market which depended on the successive interest, power and measures of four key players: the amenity societies, the City Council, the private sector, and the National government. The Netherlands case instead illustrates how the decision to reshape comprehensively the city of Groningen emerged locally in consensus and was made 'not at any one traceable moment nor by any specific local government agency or individual'. Heritage in particular was ascribed a catalytic and promotional role, and 'flagship projects' as well as the implementation of a new comprehensive plan, were local government decisions and financial responsibilities. In focus for the Swedish contribution is how shifting national and municipal policies towards the old urban built environments in general, were played out on the local scene of Haga. Regeneration and conservation measures were conceivable only when the image of the area had been converted into heritage.

The highlighting of 'identity' as a separate theme, opens up for a serious discussion of the conditions under which elements of the built environment are transformed into heritage in urban regeneration. While manœuvred by varying regulations and financing, the process here appears as deeply entangled in, and played out as, complex socio-cultural struggles over definition itself.

5 'Where the Past is Still Alive': Variation Over the Identity of Haga, in Göteborg

INGRID HOLMBERG

The line 'Where the Past is still Alive' is quoted from the 2000 Shopping and Culture guide on Haga in Göteborg.[1] Edited by Haga's Business Association, it markets the benefits of visiting the old and restored housing area, where today commercial and recreational activities are playing an important role. Departing from such contemporary commodity use of the successful results of an urban conservation project, this chapter intends to take the issue further in exploring the meanings attached to such assertions about the past as present.

During the 1960s and 1970s the strong threats from urban renewal and demolition generated the definition of the historical values of the, at the time, derelict area. The case is here used to illustrate how such historical place identity needs to be conceived as conditioned, not only by the physical fabric, by the past or by the framework for conservation measures, but moreover also by the very particular conceptualisations of the place that different actors played out in struggles over definition.

Twentieth-century Haga

The process of regeneration during the late 1970s and throughout the 1980s rendered the area of Haga a completely novel position: the area, conveniently situated close to the city centre, emerged as an important strolling area. Today any visitor to the city who goes on a sight-seeing tour or reads the Welcome-to-Göteborg folders can learn about the 'genuine charm of the old working-class area', which now has renovated facades, small cosy yards, and a street-life full of cafés, restaurants and shops. This transformation entailed, beside the rebuilding projects by private and co-operative as well as municipal builders, also the listing (*byggnadsminnesförklaring*) of about 60 buildings. However, in addition to successful conservation, it resulted in a radical socio-economic shift towards middle-class inhabitants. So far in the description, Haga displays many general characteristics and conventions of gentrifying areas.

In this case, the definition of Haga's historical values had important influence on the exception of the area from the urban renewal programme. However, a further understanding of the intense struggle and conflicts preceding this portrait, needs to take into account the ways in which the place identity was re-constructed through an intensified textualisation. This included re-definition, as well as displacement, of spatial meaning.

Meanings and Values of the Built Environment

Meanings and values of the built environment need to be understood as social matters of a moral-aesthetic kind: 'each society's 'moral order' is reflected in its particular spatial order and in the language and imagery by which that spatial order is represented' (Mills, 1992:150). While actors within the conservation and heritage sector often are occupied with the task of decoding the different historical spatial orders, the citation here calls for a consideration also of the representations of space as complexly inscribed in historical socio-cultural dependencies. Conservation and heritage actors produce representations of space and frame identity as historical values. Ultimately, such meanings are contingent upon the very 'moral order' of each society and are, hence, to be considered as historical and temporary.

In the case presented here, the heritage concerns are considered as traces of a 'textual community' (Duncan and Duncan, 1988) which share the same reading of a particular spatial segment. The 'textualisation' of space, produced by such a textual community, have a similar focus, a similar understanding of what are the important spatial properties to be emphasised and conceptualised. Understood in this way, the meaning of space becomes 'naturalised' under such circumstances. However, the complication remains that spatial meaning is under contestation and flux, as maintained by Landzelius (1999:84) 'signification with regard to space in particular, must ... be understood as fundamentally unstable and dynamic. Further, meanings and values must be understood to continuously emerge from a multitude of parallel subject positions and signifying practices'. This implies some restrictions for the understanding of 'textual community' and 'textualisation', which should not be conceived of as providing the key to a coherent 'cultural' meaning, purified from conflict, contradiction and fragmentarisation.

Also another aspect of spatial meaning has to be considered. The conceptualisations of space, apparent as coherent 'textualisations' or as verbalised meanings of a more dispersed kind, produce divisions of space through the concepts that are employed. The spatial meanings in this way tend to map out only those properties of space, to which there exists a corresponding concept. In this sense, defined historical values correspond to the physical fabric itself in particular ways, and embrace only certain spatial units. The very verbal concepts employed, apparent for example in texts, are here considered as making up a 'discursive space': an inseparable unit of both linguistic and physical prerequisite. Accordingly, the concept of discursive space denotes a construct of a word as well as its corresponding space.

In examining and re-composing some aspects of how values and meanings have been ascribed to the built environments of Haga over time, the attempt of this chapter is to explore how certain historical values make up such a discursive space. The conditions for Haga's identity, herein appears as, above all, defrayed and deeply intertwined in social concepts on other levels.

Haga in the Local Landscape

The first concern is to situate Haga in the local urban landscape. The discussion below relates materials which address real conditions in Haga. Since the aim is not to explain the occurrence of the material, but to expose some aspects of its moral-aesthetic content, the comparison or checking out with such conditions is left aside. The focus is on the traces of imposition of values and meanings (Holmberg: forthcoming).

Since the 1860s, local investigations repeatedly exposed Haga as the most miserable of areas regarding dwelling conditions. One of the most influential was a 1932 municipal investigation which concerned the conditions of the buildings, marking out the existing built environment as completely undesirable. The investigation was published some years later in the important National Investigation (SOU 1947:26), the so-called *Bostadssociala utredningen* (Folkesdotter, 1981; Rudberg, 1981). A sociological examination of Haga's residents of 1949 is here chosen to illustrate the repeatedly reinforced position of Haga in such contexts. In the processes of explaining the results, it exposes current presuppositions and conceptions.

The sociological facts, which were collected to serve as basic data for future planning of housing in the area, appear to have caused the investigators some confusion. The results showed that many residents were discontent with their apartments, but also that there was a general contentment with the area as a whole. The puzzle caused can be traced in the corrections and explanations of the results. As the results of the question 'Wishes for the future Haga' are presented, it is commented that the inhabitants in general *'certainly* [wish], that it [clearance] gets started as soon as possible', while the figure gives evidence for only some 9.3% (Karsten-Wiberg 1949:34-35, italics added). The investigation highlights further, on the one hand, that the housing stock and the housing conditions of Haga really were neglected, but on the other hand, that the residents had no joint opinion or wish for something else. The non-optional question 'Why would you like to move to a new area' is by the majority (40%) answered with 'open site close to nature', which is commented and explained as 'implying nicer and *more modern housing*' (ibid.:25, italics added). Repeatedly, the residents are ascribed aspirations, and when they fail to come, the resident's trifling demands for the future are regretted. This can be illustrated by the comment 'it was often difficult to make them look forward and reveal their wishes ... for a *new* Haga' (ibid.:51, italics added). Also in newspaper notions regarding Haga the influence of this public and official marking is reflected: it seems as if the discussants constantly felt compelled to relate their meanings to the alleged fact that 'something' had to be done (Garpenfeldt and Jacobsson 1970:43-48).

When it comes to the heritage concerns of the period, the municipal museum's 1967 inventory of Göteborg's buildings of historical value is an example of a likewise guarded position towards the area. Although it states that Haga's town plan goes back to Queen Kristina in 1677 and that the built environment as a whole makes up an interesting environment, the remark is made that the individual buildings are of a very much later date. Only very few of the buildings were selected as being of any historical value or architectural importance. This 1967 investigation, executed on commission from the City's Central Board of Administration, was to serve as a tool for classification of historical values in the parallel on-going urban renewal of Göteborg. However, only a few years later, a renewed classification by the museum graded the buildings of Haga in an almost completely different manner. It stated that 'the area as *a whole* [makes up] a very important part from a historical perspective. From the point of view of conservation the area should therefor be preserved as *a whole*' (Göteborgs Historiska Museum 1972, italics added). The historical context of this shift is the subject of the following discussion.

Urban Renewal in Göteborg: Targets, Effects and Complications

In the early 1960s, Haga was a run down inner-city area impaired by long-term physical neglect. Deterioration had come gradually during the post-World War II period, and was due to expectations of clearance: there were on-going municipal plans for enlargements of motor-traffic routes, there were prevailing regulations that hindered construction and building activities, and there were plans as well for the expansion of the central commercial district. The population declined to 6,000 residents (from 14,000 in 1920) and consisted of many elderly and poor people.

This fate was shared with many old central areas all over Sweden. Urban renewal is in a Swedish context intimately related to the political efforts, begun in the 1930s, to construct the welfare-state and to come to terms with the housing shortage. It bred national investigations which focused on housing and social reforms, such as the National Investigation mentioned above, as well as reforms of physical planning, such as the 1953 *Lex Norrmalm* which allowed for so called 'zone expropriation' that enabled the municipalities (instead of the property owners) to profit by the property appreciation which came through municipal investment.

In Göteborg a municipal urban renewal programme began in 1958 when the question of the initiation of a half-commercial slum-clearance company was raised. The company, co-owned by the city of Göteborg and some 50 shareholding private business companies, worked out a municipal programme for urban renewal. Approved by political consensus in 1962, it was to be realised through purchase, expropriation, management, emptying and demolition of the housing stock identified as in need for clearance (Pehrson et al., 1971:24-35; Garellik, 1997:12). The rebuilding should then be handed over to private, co-operative and public construction companies which had pledged themselves to renew according to the municipal plans (Schönbeck, 1994:219-220; Campanello, 1968).

Defined as target for the renewal programme were housing areas of high density, low sanitary conditions and overcrowded dwellings, generally defined as 'areas built before 1910' which referred to areas of any social kind or architectural type. However, the largest part consisted either of a certain local kind of bye-law house called *landshövdingehus*, built between 1875-1945, with two storey wooden construction on top of a brick basement floor, or of older two or three storey wooden buildings (Schulz 1988, GKH 1974:356 A: 49-63). Identifying 11 areas for renewal actions areas in the programme, several were intended to be managed by so called 'total clearance' (Pehrson, 1971:45A). Due to the effective organisation, Göteborg took the national leading position in demolition of urban buildings from the year 1965 and onwards. In total 18,361 flats disappeared between 1963 and 1972, 83% of which were demolished and the remaining part renovated (Schönbeck, 1994:217, GKH,1974:356A: 187). It is no overstatement to maintain that a large number of housing areas were cleansed of wooden buildings.

In every respect Haga was included in the intentions for total renewal. The company in the late 1960s had accomplished the acquisition of 130 properties (out of c. 300) and regarded itself to have the total domination of the market in Haga (Campanello, 1968:82). With its mixed character of buildings, but also a more complicated and complex structure of real estate property ownership, the case of expropriation in Haga in fact became a slower process than expected. By the beginning of the 1970s only about 60 buildings had been demolished (Schönbeck, 1994:219). Also a tendency to speculation by other real-estate actors could be discerned, and the company had to orient real estate business to some more remote areas (Pehrson et al, 1971:32). On top of this intricate problem, came another that was more unexpected. The plans and measures of urban renewal in Göteborg had given rise to general and vociferous dissatisfaction.

Schulz (1988:97) suggests three reasons why the debate on the effects of urban renewal was non-existent for almost ten years: a public consensus to replace existing physical structures; the labour movement's conception of the old areas as symbolising poverty and powerlessness; and the unprofitable conditions for the owners of the old properties. The proponents of urban renewal used arguments of an economic, social, technical, aesthetic and historical kind. The same arguments were gradually used also by opponents to the renewal project, such as academics and students, but as a questioning of the rationality of the whole urban renewal project *per se*. Basically it was the experience of loss which was the impartial reason for reaction to the municipal plans (ibid.: 94-98).

Regeneration Planning: Scope and Measures

This presentation of the different stages in the decisions on Haga's future will concentrate on the municipal and museum measures. Left aside in this case study are the many and important moves by the local dwellers association *Hagagruppen*, more or less associated to, or intertwined with, scholars at the local University and the School of Technology. Working hard to obtain influence in the planning process, they emphasised the local perspective and the existing working social structures and worked out alternative and competing plans for Haga.

A key player in the commencing aim to save Haga was, however, the municipal museum. Embracing the same spatial scale as the renewal programme – 'the whole' – the museum's new inventory of Haga of 1972 claimed that the whole area should be kept. It focused on larger segments: '*well-kept passages and quarters* that were left un-effected by demolition' (Göteborgs Historiska Museum, 1978:2 italics added). Also a 1973 working group on municipal basis, with the aim to develop a separate municipal Haga-programme, comprised in its scope a focus upon larger segments of well-kept properties. This approach was followed up in 1976 by the City Council decision that Haga should remain a mixed-use area with both residential and commercial activities; that the existing town-plan should be kept; and that new construction should be adjusted to existing buildings. This implied a new conception of city planning, but still no decision regarding which buildings to save.

However, a more specified focus followed in a report by the City in 1977 which classified the remaining buildings into four groups ranging from 'maintenance' to 'demolition'. This meant that 68 properties were in danger of evacuation and demolition, and already the following year demolition was permitted for 20 of them. In this, also the conservation interests were transported to another level of concern. Due to a prevalent distrust, the national system of listing buildings (designating on the level of the British Grade 1, however with stronger requirements) was being used by individuals in the hope to rescue them from demolition (see chapter 3). In 1979 as many as 60 buildings were proposed for such listing. The County Administrative Board and the National Heritage Board then took the initiative to develop a conservation plan, or programme, to guide the listing activities. The programme required an assessment of the historical value of each separate building, regardless of its site. The completed programme of 1978 eventually covered two complementary categories: A) 24 'unique and characteristic properties' (well kept), and B) 39 complementary 'characteristic' properties (well kept).

What distinguished the regeneration of Haga from other Swedish cases, was the extensive national support obtained. In 1979 the National Government decided to increase the level of special supplementary straight loans, free of interest, to enable the restoration of buildings of 'great historical value'. This meant that for the first time supplementary loans – the level of which was raised up to 100% – could be used for the costs of the adaptation of construction techniques to the requirement of the old buildings. The pool was administered centrally at the National Heritage Board, but the responsibility for deciding on the individual cases was on the level of the County Administrative Board. In 1979 it was centrally decided by the National Heritage Board to offer the major part of the pool of these loans to the buildings in Haga proposed for listing in the programme. Also, in the same year, special grants for the adjustment of new construction to the existing building structures were admitted by the National Housing Board (Andersson, 1985).

The administration of the Swedish listing system is in fact dependent on the owner of the building accepting the proposal. The city of Göteborg, being the owner of most of the properties, delayed the decision on the proposals for listing for two years, and then, in February 1981, demanded a more flexible listing programme, enabling a closer connection of objects. Seven properties were then left out of the programme, and another three properties were exchanged. This was however not enough. The financing of the re-building and restoration projects in Haga needed an estimated sum of 20.3 million SEK, and the Real Estate Council required further adjustments in the programme for continued financing: eight properties to be deleted from the programme, and another six properties to be added. An acceptance with some conditions followed from the National Heritage Board and the County Administrative Board in November 1984, and later the same year, the City Council of Göteborg eventually decided to accept the listing of 60 buildings in Haga.

Nineteenth-century Haga in Construction

Realisation of 'Social Time': Conditions for Identity

It is obvious that the shifting selection of buildings, as well as larger segments according to different priorities, has been caused by a struggle over the definition of Haga's identity. It may also be observed that aspects of identity that concerned the historical value could be an object for negotiation. Exploring the content of the historical values, it is, hence, necessary to turn to a discursive level of construction.

The arguments against urban renewal were principally presented as 'new knowledge' which can indeed, in Foucauldian terms, be understood as acts of 'counter-archaeology of social knowledge', giving rise to new perspectives, but moreover, counter-acting on the very same discursive arena as the hitherto dominating. The aesthetic and historical arguments are in this context the most rewarding to observe since they were intricately entangled in the conservation concerns.

The criteria for selection are one such concern. It has been observed that during the period traditional criteria for selection of the 'historically valuable'– rewarding the 'unique' or the 'artistic' – were replaced by a 'neglected history of the people' (Schulz, 1988:97). Such a controversy over historical meaning can be understood as the reason for the new efforts to write the history of these areas in order to make visible a conceived repressed past. Herzfeld (1991:10ff) adds a temporal dimension to the understanding of processes of reclaiming the past and conceives them as realisations of different conceptions of time: the official, nationally bound, 'monumental', reductive and generic in its character, imposing order; versus the 'social' which is inclusive and unpredictable, in the grist of everyday experience, and which takes into account the local experience of daily practices.

It can be suggested that in the case of Haga, the attempts by conservation seems to have been to embrace the historical values of the urban built environments as exponents of 'social' time. This was a break with earlier measures and required a different approach and terminology than usually used for discerning a past of a 'monumental' kind.

The following passage will introduce some of the modes in which conservation also attempted to embrace aspects of the different pasts and different spaces of 'social time'. This has bearing on the issue of aesthetics, and indicates the intricate linking, within practices of conservation, of conceptions of historical value to an aesthetic organisation. This leads to another aim of this chapter, to shed light on the conceptualisation of Haga's historical values as relating to space in particular ways and making up a 'discursive space', the holders of which will be explored.

'The First Suburb of Göteborg and its First Workers' District'

The attempt to inscribe 'social time' in conservation was accomplished through the ascription of values and meanings which before had only infrequently appeared in the classifications of historical values. One such realisation of 'social time', is the emphasising of the worker as a vehicle of meanings and values. While the position of Haga in the moral landscape of Göteborg for a long time concerned the bad physical and social conditions of the area, the label 'worker' appeared as a way to render Haga a solid identity and place in history, exemplified as in the header above drawn from the guide-booklet on Haga by the municipal museum (Lönnroth, 1979). Seen in this way, 'worker' was a way to orient the search for the historical value towards aspects of the 'social time' of the lived place of Haga. I will forego further discussion of the conceptions attached to the term 'worker', but will in the following simply discuss some possible implications for the conservation project.

Only since the 1970s has the central theme in the conception of Haga been that it essentially is a workers' district. It is interesting, that it is absent in a 1970 publication which comprises an extensive historical overview over Haga's buildings and people (Kjellin, 1970). Here Haga is a 'suburb' or an 'area'. Another approach is prevalent in a guide from 1978 edited by a member of the local dwellers association *Hagagruppen*. Asking 'Why is it that we fight for saving an old area with outside privies, peeling facades and vacant demolition sites?' the answer is that 'Because Haga is Göteborgs's oldest workers' district' (Hansson, 1978:3).

The labelling of Haga as a 'workers' area is here appearing as a rhetoric, but an analysis of the labelling brings some complications. Although the municipal museum was one of the major actors in rendering Haga a 'workers' identity, passages in the museum's guidebook can give examples of contrasting and contradictory facts about the nature of Haga's past. The booklet expounds on Haga's origin as a suburb to house workers, but it seems that in no period did Haga only housed workers or poor people: one learns that beside the workers of diverse kinds, there were 'merchants', 'crafts-men', owners of the means of production in Haga's factories, 'inn-keepers', 'house-owners', and 'grocers'. The booklet's headings nevertheless run: 'The Boatswains and the Mast-cutters' (seventeenth century), 'The Iron-carriers and the Labourers' (1700-1840), 'The Factory Workers and the Seamstresses' (1875-1930), which reinforces the genuine 'worker' image at every instance (Lönnroth, 1990). The facts delivered in the booklet regarding social and physical complexity, such as lavish buildings and wealthy groups, hence, confuse and contradict the homogenised labelling of Haga as a workers' area, and instead introduce a more complex character which is fundamentally socio-spatially mixed.

'Workers' area' is here singled out as one 'discursive space' giving Haga a firm identity. If social 'mix' is invalid for the generation of identity, this fact demands an expounding on what would be the content of an ideal workers' identity. The question is what spatial properties are conceived as applying for this concept. I will exemplify from the rhetorics of selection, which I argue can be analysed with two foci: one is the all-encompassing conception, which on a discursive level is contingent upon a relationship of 'worker' and 'all'. The other concerns the restricted conception, which is apparent in the discursive exclusion of particular physical structures (as well as in emphasises of certain other). This offers an opportunity for consideration of aspects of the core, possible limitations as well as dynamics, of Haga's worker identity.

The core of the worker appears in the museum's guide-book. It suggests to the visitor a guided tour that starts at an outer surrounding alley: 'Departing from the lavish stone buildings out here, you can make your way into the *more simple* and for the area *more characteristic* wooden blocks' (Lönnroth, 1979, italics added). In 1972 the stone buildings made up about 30% of the buildings in Haga, back in 1920 the proportion was about 25% (Göteborgs Historiska Museum, 1972: Appendix 2). But also the 1978 museum's examination of the buildings declares how 'roughly all the stone buildings erected after 1880 ... as well as the *landshövdingehus* built after 1910, are left out of the programme' (Göteborgs Historiska Museum, 1978:4). Also this example gives prominence for wooden constructions over those of stone material. The core of the discursive space of the worker is wooden.

When it comes to the restriction and limits of the discursive space of the worker, an illustration can be drawn from a statement which declares that 'Characteristic for Haga is, above all, the *mixture* of wooden buildings, *landshövdingehus* and stone [brick] buildings together with the old city plan' (Göteborgs Historiska Museum, undated, italics added). This characterisation of the physical properties can be understood as contradictory to the statement above of the 'wooden' as ideal type. More over, in every selection of historically valuable objects made by the museum, several stone-buildings are included. The limitations of the discursive space, hence, have to be searched for on another level. For example: not visible in the citation is that parts of the *outer* stone building are high-rise, while the *inner* are low-rise. Adding this physical property, 'high-rise' can be considered as one such limit of the discursive worker.

Examples drawn from another spatial scale suggest a more detailed concern to safeguard particular looks conceived as necessary for the worker identity. In a presentation of the ongoing conservation measures after the final decision of some 60 buildings for listing, a museums' officer introduces the over-all task as an attempt to 'maintain the *impression* of a *worker's area* from the [last] turn of the century, so that there remains a possibility to imagine aspects of the *hard and poor life* that was lived here'. Accordingly, the restoration was not to become 'too idyllic' (Lönnroth and Tengnér 1985 p. 27). The back yards, being 'simple, and often quite harsh' spaces, were also emphasised as spatial sections of particular concern (ibid. p. 28). In some cases this required additions for the right look – such as chimney dummies – while in some cases this resulted in restrictions towards additions of functions – such as a new mechanical refuse system in the backyard of a residential building where the objection was noise. Regarding the exterior colour scheme '[usually] the original colour has not been compiled', but instead, 'it has been a goal to use colour to *reinforce the impression of a workers' area*' (ibid. p. 30, italics added).

Conclusion

Setting out from the notion that historical value is a particular conception of place, inscribed in complex socio-cultural and politico-economical dependencies, and that any such conceived place identity also entails a social content with moral-aesthetic implications, I have tried to enfold aspects of one such vehicle of social meanings which departs from the 'workers' area' identity of Haga.

This identity appeared when a radicalised conservation movement broadened its scope to embrace also the 'social time' of the everyday vernacular. From the examples, an ideal 'worker' realising 'social' time, however, primarily seems to be an aesthetic project. To be observed is here, how *primary functions* of the restored spaces – possible to associate with the conditions of the turn-of-the-century workers' dwelling, such as noise from functional systems – were less included in the sign vehicle of the discursive worker, than were the concerns for *secondary functions* like certain colour-schemes or ubiquitous aesthetic disharmony of absent chimney symmetries. It remains to be seen if the discursive space of the worker – as verbal and spatial constant – will remain as a future superior order for interpretation of Haga.

Note

1. The citation is a translation from a header with the Swedish wording '*det gamla*'. In the English summary of the guide, the line runs 'Haga, where *history* is still alive' (italic added) which does not sufficiently capture the original sense of the Swedish words.

References

Andersson, H. (1985), 'Haga', *Kulturminnesvård* vol. 6 pp. 3-7.
Campanello, L. (1968), 'Riva för att Bygga', in *Göteborg Förr och Nu*, Göteborgs Hembygdsförbunds skriftserie V., Göteborg.
Duncan, J. and Duncan, N. (1988), '(Re-)reading the landscape', *Environment and Planning D: Society and Space* vol. 6 pp. 117-26.
Folkesdotter, G. (1981), "Störtas skall det gamla snart i gruset" *Bostadssociala utredningens syn på äldre bebyggelse*, Meddelande M81:12 Statens Institut för Byggnadsforskning, Gävle.
Företagarna i Haga (Haga Business Association) (2000), *Haga. Shopping och kultur*.
Garellik, R. (1997) *Göteborg före grävskoporna. Ett bildverk*. Printed in Borås.

Garpenfeldt, B. and Jacobsson, H. (1970), *Upprustning av Haga. Ett saneringsalternativ*, Department of Architecture, Graduation thesis 1970:3. Chalmers School of Technology, Göteborg.

Göteborgs Historiska Museum (1967) Kulturhistoriskt värdefull bebyggelse i Göteborg. Inventering.

Göteborgs Historiska Museum (1972), *Haga – Bebyggelsehistorisk utredning.*

Göteborgs Historiska Museum/ Länsstyrelsen i Göteborgs och Bohus län (1978) *Utredning angående byggnadsminnen i Haga 1978.*

Göteborgs Historiska Museum (undated) Kulturhistoriska synpunkter på Hagas sanering. Historiska museets bedömning av Haga.

GKH Göteborgs Kommunfullmäktiges Handlingar 1974:356 A: *Betänkande med förslag till riktlinjer och program.*

Hansson, E. (1978), Vägvisare över Haga. En rundtur i Göteborgs äldsta trähusstadsdel – en levande miljö som nu hotas av rivning. Korpens förlag Göteborg.

Herzfeld, M. (1991), *A Place in History. Social and Monumental Time in a Cretan Town.* Princeton University Press, New Jersey.

Holmberg, I. (forthcoming) *Spaces of the Past/the Past as Space*, Diss. Institute of Conservation, Göteborg University.

Karsten-Wiberg, E. (1949) *Sociologisk undersökning i Haga våren 1949*, Redogörelse utarbetad å Göteborgs stads statistiska byrå.

Kjellin, M. (1971), *Haga i Göteborg*. Fastighetsaktiebolaget Göta Lejon, Göteborg.

Landzelius, M. (1999), *Dis[re]membering Spaces, Swedish Modernism in Law Courts Controversy*, Diss. Institute of Conservation, Göteborg University.

Länsstyrelsen i Göteborgs och Bohus län (1992) *Göteborg. Kulturmiljöer av riksintresse*, Printed in Göteborg.

Lönnroth, G. (1979), *Haga. Göteborgs första förstad och arbetarestadsdel*, Göteborgs Historiska Museum, Göteborgs Stadbyggnadskontor.

Lönnroth, G. (1990), *Haga. Göteborgs första förstad och arbetarestadsdel*, Göteborgs museer.

Lönnroth, G. and Tengnér, L. (1985), 'Antikvariska problem', *Kulturminnesvård* vol. 6 pp. 27-30.

Mills, C. (1992), 'Myths and Meanings of Gentrification', in Duncan, J. and Ley, D. (1993) *Place/Culture/Representation*, Routledge, London and New York.

Pehrson, G., Ranch, C. and Sabel, A. (1971), *Göta Lejons betydelse för saneringsverksamhet i Göteborg 1960–1970. En analys och värdering*, Företagsekonomiska studier. Göteborgs universitet. Rapport 1971:43. Göteborg: Gothenburg Studies in Business Administration.

Rudberg, E. (1981), *Uno Åhrén: en föregångsman inom 1900-taletsarkitektur och samhällsplanering*, Diss. Stockholm: Tekn. högsk. Stockholm: Statens råd för Byggnadsforskning. Svensk Byggtjänst (distr.). Serie: T / Statens råd för byggnadsforskning 1981:11.

Schulz, S. (1988), 'Argument i den göteborgska stadsförnyelsedebatten', *Årsbok 1988 Arkitekturmuseet.* Stockholm.

Schönbeck, B. (1994), *Stad i förvandling. Uppbyggnadsepoker och rivningar i svenska städer från industrialismens början till idag.* Byggforskningsrådet T16:1994, Stockholm.

SOU 1947:26, *Slutbetänkande avgivet av Bostadssociala utredningen*, Del II.

6 Nottingham Lace Market

GRAHAM BLACK

Nottingham Lace Market was designated a conservation area in 1969, one of the first batch of nine in the City and the first industrial zone in the country to be so designated. This case study brings together the results of research on the Lace Market in an attempt to enhance understanding of the role of conservation area status in its substantial revitalisation over the last thirty years. The study also seeks to highlight issues being addressed today which are vital to the area's future, particularly the relevance of the area's history to its occupants today. Is the area's heritage value now only skin deep? Has the role of conservation area status really been about regeneration and creating a new future?

The Townscape of the Lace Market

Today, the Lace Market is a remarkable area, a nineteenth century island situated on the south-east fringe of Nottingham's city centre. Its location, astride what was once called St Mary's Hill, gives it a prominence in the City rivalled only by Nottingham Castle. As the former nerve centre of the machine lace industry, in a city which was the global heart of the industry in the nineteenth century, its strong identity and sense of place comes from a combination of grand four and five storey Victorian warehouses superimposed on a pattern of surviving, narrow Saxon and medieval streets. This combination gives a canyon-like appearance and feel to many of the streets (see Figure 6.1).

The majority of the warehouses were erected between 1853-1906 and many are of individual architectural importance. They were designed mainly by Nottingham architects. The buildings are constructed mainly of red brick, with heavy use of stone at ground floor level and for the large windows (good light was essential for the tasks associated with 'finishing' the lace). Their scale and frequently ornate decoration reflect the wealth and confidence of their builders. The area as a whole is of outstanding national and international significance as an example of a concentration of industrial buildings connected with one trade.

74 The Construction of Built Heritage

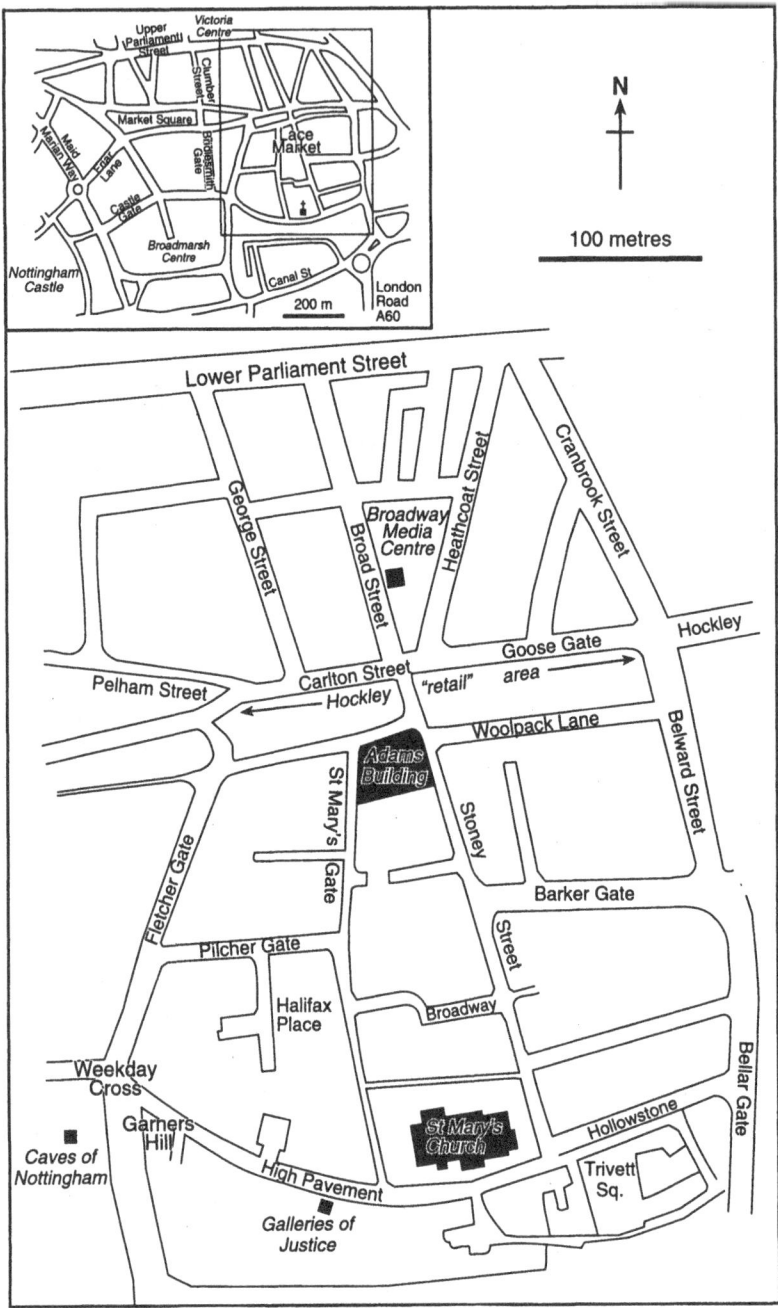

Figure 6.1 The Lace Market Area, Nottingham
Source: Reproduced from the Ordnance Survey based mapping by permission of Ordnance Survey on behalf of the Controller of Her Majesty's Stationery Office, © Crown Copyright ED 100017895

The Rise and Fall of the Lace Market

The Lace Market area occupies a naturally defensive location, rising above the valley of the river Trent. The first settlement of Nottingham grew up to its east, but was moved to this area in the late ninth or tenth century. The heart of the settlement then gravitated further westwards after the building of the first Nottingham Castle in 1068. Medieval Nottingham was in decline by the fourteenth century. The earliest map of the town (1609) shows most of its east side - that is, the old Saxon town and future Lace Market area - largely vacant. Here the gentry built town houses in the seventeenth and eighteenth centuries, to be replaced later initially by workers' housing and then by lace warehouses.

The Nottingham lace industry grew out of experiments by workers in an already established hosiery industry. The first hand-operated machines produced a type of plain lace net. By the 1840s sophisticated steam-powered machines could produce all types of patterned lace. By the 1860s machine-made lace was hardly distinguishable from the hand-made variety, yet was produced at a tiny fraction of the cost.

The lace machines were located in factories away from the town centre. However, the industry was dominated by merchants who acquired the lace from the machinists, co-ordinated the almost forty processes required in 'finishing' the lace (preparing it for sale), and then found buyers. Many aspects of the finishing process were inter-related, so the merchants tended to congregate together. By the 1830s their greatest concentration was in the Lace Market area, and the area had already been given this title by 1850.

Initially merchants operated from converted private houses or small general purpose warehouses. The 1850s saw a massive increase in scale, with new buildings designed both to impress potential trade buyers from all over the world and to provide the right conditions for lace 'finishing'. The term 'warehouse' is itself misleading. The buildings were not only used for storing and despatching lace, but also for some of the 'finishing' processes and as the showrooms and commercial sales centres for the lace companies. The first new warehouses, opened in 1855, were erected on vacant land or, in the case of Broadway, on land recovered from an abandoned town mansion. More land rapidly became available, however, as new workers' housing was built away from the town centre. By 1910, when the last warehouse in the area was completed, the Lace Market had the appearance which it still largely retains today.

The industry fell into steep decline after the First World War, as fashions changed and competition from France and Germany grew. In addition, the surviving industry began its move away from the city centre (Nottingham was made a city in 1895). The great warehouses emptied rapidly or were sub-divided for multiple use, particularly in the making-up trade for the clothing industry. As early as 1936 the Nottingham Journal could write:

> The best thing that could happen is for the ground to be cleared and a fresh start made.
> (Nottingham Journal Trade Review 1st January 1936)

In 1950 the Adams Company, symbolic of the industry at its peak, ceased trading and the Adams building, the most significant warehouse in the Lace Market, was itself sub-divided. There was no great admiration of Victorian architecture in the UK in the 1950s. Nottingham's first Development Plan, published in 1952, makes no mention of the quality of the Lace Market area or its economic problems.

As the lace industry contracted from the 1920s, and surviving companies moved their activities away from the Lace Market, much of the vacant space was occupied by a developing clothing industry. An industrial survey of the area in 1969 showed almost 6,000 people were employed there, 79% in manufacturing. Some 90% of this 79% were employed in textiles and clothing (Smith, 1985). Like the lace industry before them, many of the clothing companies were inter-related, supplying goods and services to each other. They also needed a city centre location to attract their highly-skilled, largely female staff. However the rental the textile companies were willing or able to pay was very low, insufficient to maintain the buildings, let alone restore them. The overall result for the area was a feeling of physical decay.

Apart from the area's general economic decline and resultant atmosphere of dereliction, the greatest threat to its physical survival came not from the private sector, but from the City Council itself, with a plan to create an inner ring road, published in 1965 as *Traffic in Nottingham 1965-2005*. This document, never even submitted for approval to the full City Council or to central government, was nevertheless acted upon by the then Planning Committee as if it was official policy; it was finally rejected by the relevant government minister in 1971, following an appeal by the City's fledgling Civic Society. For the Lace Market area, it resulted in the demolition, in the mid-1960s, of a number of warehouses along what was to have been the route of a feeder road.

Revival: the Legislative Framework

Despite its dilapidated condition in the mid 1960s, the Lace Market contained the highest concentration of listed buildings within the City - although many of these were examples of Georgian domestic architecture located on High Pavement, on the southern edge of the Lace Market. However, this offered no protection to other properties or to the overall townscape. The breakthrough came with the Civic Amenities Act 1967, following which the City Council designated the Lace Market as one of Nottingham's first batch of nine Conservation Areas, in 1969.

However, the action required to revive the area depended on more than specific conservation legislation. The provision of grant aid to support refurbishment began under Section 10 of the Town and Country Planning Amendment Act 1972, and has continued ever since in a variety of guises, under such legislation as the Inner Urban Areas Act 1978, the support of various temporary employment programmes, funding from the European Union and even the establishment of the National Lottery. There is no need here to define the precise nature of the individual pieces of legislation - what is relevant is the effectiveness of the City Council, over a sustained period of time, in gaining and applying grant aid from a wide variety of sources.

Revival: Players and Actions

If Nottingham City Council was the chief villain of the piece until the mid-1960s, it has since become the Lace Market's guardian angel. However, it could never have succeeded alone. What is fascinating is to explore the range of players involved, how they made use of the legislative framework and how their role has changed over time. The first key players were the amenity societies. In 1965, the Council for British Archaeology published a list of 324 towns, 'the historic quality of which particularly required treatment in any planning or redevelopment proposals'. Nottingham was included in the list because of the threat posed by redevelopment to its archaeology, because of its ancient street plan, and because of the quality of its Victorian buildings. Attitudes to nineteenth century architecture were changing, not least because of the activities of the Victorian Society, founded in 1958. There was also increasing public concern over the rate and large-scale nature of redevelopment in city centres. In Nottingham, a newly created local Civic Society began to campaign against the worst proposals - not least the proposed road scheme referred to above - and to promote the City's Victorian buildings.

The City Council's change to an actively supportive role, following the designation of the Lace Market as a conservation area, did not just apply to an appreciation of the area's architectural quality. It also included an acceptance of the positive role of the amenity groups. The Council established a Conservation Areas Advisory Committee including councillors and council officers, but also members of relevant local amenity societies. A working party was set up to work out detailed proposals for the conservation of the Lace Market. It considered and rejected both wide-scale redevelopment of the area and allowing it to continue to slowly decay. When the working party reported in 1973, it came down in favour of taking positive measures both to preserve the economic viability of the area and to encourage inward investment to give new life to the buildings (Smith 1985). The City Council adopted this report and also agreed to extend the existing conservation area north from Goose Gate to Parliament Street. This was the 'housing and social provision zone', north of the warehouse complex, where many of the lace workers had lived and where buildings which served their educational, religious and social needs were located.

Nevertheless, even working together, the public sector and the amenity societies could not guarantee the future of the Lace Market. The concept of a Conservation Area has always contained a basic dilemma. The public sector might consider the area so important that it should be designated, but it would never have the financial resources required to carry out necessary maintenance and improvements. That depended on securing a viable future for the buildings, and made a third player - the private sector - an essential part of any solution. In reality, it made the private sector the lead player in the Lace Market's revival. The majority of Conservation Areas in the UK consist substantially of privately owned domestic properties. Owners in these areas tend to support Conservation Area status, with occasional exceptions, because it has proved to enhance the value of their properties. The Lace Market, as a commercial area, was different to this. Commercial property owners had to be convinced that they could make a proper financial return through renovation of the buildings rather than demolition and rebuilding. Until the national economic recession of the late 1980s, renovation of the warehouses led inevitably to their conversion to office use which, during the 1980s property boom, could attract five times the rental income of their previous use by the textile industry.

The Council's initial approach was to serve repair notices on the owners of listed buildings in the area and to resolve to acquire property where necessary to ensure preservation. There was also a recognised need for environmental improvements to enhance the area as a whole. Most importantly, a means had to be found to encourage owners to restore their buildings. The solution, reached gradually at a national level - and so bringing national government into the equation as the fourth player - was to make grants available from public funds through schemes which would persuade building owners to carry out necessary work on their buildings, help to bridge the perceived gap in profitability between restoration and new build; and potentially convince owners to repair the buildings to a higher standard than they would have considered without grant aid. These schemes involved what at the time was a new concept of co-operation between central government, local government and developers.

The most important source of grant aid was the 'Town Scheme', introduced in 1976 when the Department of the Environment declared the Lace Market to be of outstanding national importance. The scheme was administered initially by the Department of the Environment then, from 1st April 1984, by the newly-created English Heritage. The first Town Scheme in the country was the Bath Terraces Scheme, set up in 1955. However it was only in the 1970s that the scheme, to grant aid conservation activity in towns, spread nationally. In the Lace Market, grant aid was given not only for the restoration of the lace warehouses at the core of the conservation area but also for the refurbishment of small retail premises in the Hockley area, on the northern fringe of the warehouse complex, bringing new life to this key zone. A further dimension was added in 1978, when the Department of the Environment introduced 'Operation Clean-up' to improve the visual environment of cities like Nottingham, on condition that eligible local authorities met 25% of the costs. An important aspect of the scheme was grant aid for environmental improvements to derelict sites. This work was also supported by temporary employment schemes. By the early 1980s, around twenty derelict sites in the Lace Market had been landscaped or otherwise tidied.

Meanwhile, a 1977 government white paper, *Policy for the Inner Cities*, had pointed to the need to support the economic well-being and community life of inner urban areas. It expressed concerns, for example, about the under-utilisation of buildings and that a shabby environment was seen to be a disincentive to those wishing to invest in business and industry. Under the resulting legislation, the Inner Urban Area Act 1978, Nottingham was able to prepare an annual rolling programme to combat inner urban decay and to declare Industrial Improvement Areas. In 1979, the Lace Market was declared an Industrial Improvement Area, with property conversions eligible for up to 50% grant aid.

By 1982 more than one hundred buildings in the area had been renovated (Crewe and Hall-Taylor, 1991). In addition, by the early 1980s, three large vacant sites were developed for new housing, most importantly a housing association project at Halifax Place, a site which had been derelict for forty years. This latter project reflects another role for amenity groups. The housing association involved had been established by members of Nottingham Civic Society. Their ambition was twofold - to create high quality new-build which would blend in with the surrounding Victorian architecture, and to inject life back into the area through the sixty-four new dwellings involved. Again, grant aid was involved, this time from central government's Housing Corporation. In 1983, the scheme won a Housing Design Award.

In 1983 Nottingham City Council was given the Royal Town Planning Institute's Silver Jubilee Cup for its work in the Lace Market and also received a coveted *Europa Nostra* award. Work increased in pace during the remainder of the 1980s, reflecting a national economic boom then under way. By 1988 it was the common view that the Lace Market area had been transformed; the quality of its architectural heritage has been realised, and, on the surface, its future appeared to be secure.

However, recession from the late 1980s to the mid 1990s showed that a headlong rush into office conversion could not, on its own, provide a sustainable future. Because of the nature and condition of the buildings, it was not commercially or architecturally viable to covert them to first rank office space, with full air conditioning, false floors and ceilings for computer related wiring, etc. Not surprisingly, these offices were among the first to be vacated when recession hit. By 1993, at least 15,000 square metres of office floor-space in the Lace Market was vacant, representing almost 30% of total vacant city centre office floor space. A further c. 25,000 square metres of new build and conversions had not advanced beyond planning stage (Nottingham City Council 1993).

Also, central government grant aid was being phased out. In 1993, the Lace Market was not included in the 'Conservation Area Partnerships' with which English Heritage replaced the old Town Schemes. Equally, Industrial Improvement Areas were replaced by a new scheme, 'City Challenge', and the Lace Market again failed to obtain funding. The view must have been that the Lace Market had received enough support and that there were other areas in greater need. Their role has since been replaced by new financial players, the Heritage Lottery Fund, whose priority lies with the heritage quality of the proposed project, and the European

Community, with a priority of grant aid in return for economic benefit. However, there is far less of a guarantee of funding from these sources than from the rolling programme of central government resources they replaced.

In 1989, at the tail end of the 1980s property boom, a Lace Market Development Company was created as a partnership between Nottingham City and Nottinghamshire County Councils and a group of private property developers, as a mechanism for maintaining the scale of redevelopment in the area. However, this coincided with economic recession and a property-led initiative proved unsuitable. The Company still exists, but operates on a very limited scale. However, one of its key initiatives was the creation of a Lace Market Heritage Trust, formed in 1991 by representatives of a range of Lace Market organisations committed to seeing the area used to its best advantage. With the active support of the City and County Councils, the Trust has become a major new player in the area. Its activities began with the development of a Lace Market logo and the part funding of new street signs and street furniture. It has since become a key agent in the enhancement of the area for visitor use and played a vital enabling role in the funding of the conversion of the Adams Building for educational use (see below).

During the 1990s the City Council commissioned a series of reports on aspects of the future direction of the Lace Market:

- Tibbalds / Colbourne / Karski / Williams: *Nottingham Heritage Area Study*, 1991.
- Black, G and James, A: *The Lace Market Nottingham, A Vision and Strategy for Visitor Use*, 1992.
- Crewe, L.J. and Haines, L: *Building a civilised and competitive city: The Lace Market as a cultural quarter*, 1996.

These were used as a basis for its own *Lace Market Strategy Reviews*, carried out in 1993 and 1996. These recognised that no single initiative would be adequate. There was a need for an integrated strategy involving planning, design, interpretation and marketing promotion. They also continued to give priority to retaining the textile industry in the Lace Market. As a result, in 1993 the City Council approved a long-term programme of environmental improvements, with an estimated cost of £2.25m, including improving pedestrian links with the city centre and Castle Museum area; creating a sense of arrival at the gateways to the Lace Market; upgrading the Hockley and Stoney Street areas; encouraging pedestrian activity; improving access for the mobility impaired; and enhancing the Lace Market's identity.

City Council support for the development of visitor use in the area was essential to the work of the Lace Market Heritage Trust in raising over £10m - including funds from the Heritage Lottery Fund and the European Commission - between 1993-1999, to create the Galleries of Justice, a Museum of Law, in the historic Shire Hall on High Pavement. This has since won numerous national awards. The City also part-funded the opening, as an attraction, of an historic cave complex located below the adjacent Broadmarsh shopping centre. It was hoped that these new developments would in turn encourage private sector investment in adjacent facilities (cáfes, gift and antique shops etc.) that would convert the area into a true visitor destination.

However, other factors have had a much greater impact.

Students: the conversion of the Adams Building on Stoney Street for use by New College, a local college of further education, has brought some 8,500 students each year into the area, creating jobs in its own right, enhancing the local skills base and having a major knock-on effect in terms of student spending in the area. It opened to students in late September 1998. This £16m project received substantial funding from both the Heritage Lottery Fund and from the European Community. It reflects a further conservation issue in that, as the largest and most important building in the area, it could not attract a viable new use until massive grant aid was made available for a public sector scheme. The role of the Lace Market Heritage Trust in facilitating the scheme reflects again on the effectiveness of public-private partnerships.

Accommodation: the potential for additional residential accommodation in the Lace Market was finally revealed in 1999. This reflects a growing national trend to bring people, particularly the young and those without cars or children, back to live in city centres. Their presence and spending power has brought additional vitality to the Lace Market and created additional support for local shops and leisure facilities. Since 1999 there has been a rush of conversions of warehouses by private developers to create top of the range 'loft-style' accommodation, attracting the fashionable wealthy into the area. Their popularity saw prices double in the first twelve months. The area has also seen the conversion, by Nottingham Community Housing Association; of former lace warehouses around Trivett Square, off Hollowstone, to create 139 flats, mostly for rent.

The Future

It can be argued that the recession of the 1990s had a positive side to it, so far as the long-term viability of the Lace Market was concerned. It encouraged the City Council to take stock of achievements and plan for the future. It provided an opportunity to explore alternative uses, particularly education, housing, leisure, retail and tourism. It has proven to be in the interests of both the private and the public sector to achieve a more balanced approach to development rather than concentrating solely on the creation of office space. It has seen the emergence of a highly effective new player, in the Lace Market Heritage Trust, and of new sources of funding. There is real potential for this revitalisation to continue. The presence of New College on Stoney Street has added a new dimension to the area and a potential all-year-round market for shops, cafés and pubs. The conversion of former lace warehouses to living accommodation continues at a rapid pace and new residents will further increase demand for other facilities.

However, it is not all good news. The lace and textile industry has virtually disappeared. The post 1995 economic upturn has seen some of the empty office space re-occupied and much more converted to accommodation, but some continues to remain vacant. There is little pressure for further office conversions and key developments approved in the early 1990s have not progressed beyond the drawing board. Visitor use of the area has not increased as hoped and has had little positive impact in terms of job creation or private sector developments. One factor which the Black and James (1992), report failed to appreciate has, however, proven to be an outstanding success - the importance of club-life in the area. Today, the Lace Market is at the hub of Nottingham's club scene - the most important in England outside London and Manchester - and, literally, throbs late into the night.

Designation as a Conservation Area resulted in a long, hard and expensive slog to reach the point that the area is at today, and public sector involvement must continue for at least the foreseeable future. However, today, the Lace Market is seen as the 'place to be' - from trendy offices for architects and advertising agencies, to the delights of urban loft living and late night clubbing. The *present* of the Lace Market still nods in the direction of its history but it is the future which matters to almost all those who occupy it.

Conclusion: Conservation Versus Regeneration

Since its designation as a Conservation Area in 1969, the sustained actions of a range of key players has resulted in the sustainable regeneration of Nottingham's Lace Market, rather than the slow decay and eventual demolition which threatened. However, in the process, the initial objective of retaining the area's original function as well as its architecture has been lost. The architecture of the Lace Market remains a potent symbol of Nottingham's past glory as the lace capital of the world. However, although a remarkable 'sense of place' is retained because of the powerful buildings and narrow streets, there is no longer any feeling of a national textile centre. The area's identification with the textile industry is skin deep. The structures remain but what a visitor to the area now encounters has little to do with its original use and much more to do with the remarkable capacity of historic industrial buildings to be adapted to new uses. In the circumstances, given the dominance of economic issues, it is difficult to see how this could have been prevented.

From the beginning of its intervention, it was City Council policy to retain a strong textile industry in the area. In heritage conservation terms, this policy was also important, recognising that the best use for a building is normally as near as possible to that for which it was originally developed - the conservation of a living heritage, not just a built environment. However, the necessary reliance on private investment, motivated by expectations of profit, placed considerable pressure on the Lace Market's role as a textile centre. First, office rentals achieved up to five times that of textile use while, latterly, substantial profits have been made from conversion to residential accommodation.

The pressure was eased slightly by the City acquiring and converting some buildings as textile workshops and also establishing a Nottingham Fashion Centre (closed in 2000). The situation was made worse as planning controls were weakened due to central government policy. However, as late as 1991-1993, a survey showed 108 textile firms including designers, manufacturers and retailers, together employing some 1,300 people (Crewe, 1994). Most employed fewer than ten people producing high-quality exclusive design-based garments. The Hockley area had developed as Nottingham's unique fashion area, providing a contrast to the normal High Street chain stores. Since then, the continuing decline of the British textile industry, in the face of cheaper foreign competition, has led to the near disappearance of textile manufacturing from the area, although the Hockley retail area retains its identity as a fashion district.

By the 1990s, when the Tibbalds report and City Council strategy reviews highlighted the need to enhance the Lace Market's identity they were speaking not of developing new textile uses but of themed environmental improvements. It is hard to resist seeing this approach as a deliberate attempt to use the area's 'sense of place' as a resource to be exploited in development terms rather than a strategy led solely by the principles of conservation. Does this matter? Should we not be talking in terms of regeneration rather than conservation? Many of these warehouses were built to the highest architectural standards of their day. It seems appropriate, in both commonsense and business terms, to revitalize them with new uses suitable to the twenty-first century. They provide an attractive atmosphere for living and working. Their occupants are also enlivening Nottingham city centre as a whole. From being a 1960s symbol of the inexorable decline of the British textile industry, the Lace Market has been transformed to represent the most up-to-date central government strategy for re-occupying the city centres of the UK.

References

Black, G. and James, A. (1992), *The Lace Market Nottingham, A Vision and Strategy for Visitor Use*, Unpublished report to Nottingham City Council.

Council for British Archaeology (1965), *Report No 15 for the year ended 30th June 1965*. Council for British Archaeology, London.

Crewe, L. and Hall-Taylor, M. (1991) 'The restructuring of the Nottingham Lace Market: industrial relic or new urban model?', *The East Midland Geographer*, vol. 14 pp. 14-30.

Crewe, L.J. (1994), 'Consuming Landscapes: Designing in the Nottingham Lace Market'. *The East Midland Geographer*, vol. 17 1 and 2 pp. 22-27.

Nottingham City Council (1993), *Lace Market Development Strategy Review*. Unpublished report.

Smith, R. (1985) 'Issues in Urban Industrial Conservation: The Nottingham Lace Market'. *Industrial Archaeology Review*, vol. 7 2 pp. 139-153.

Tibbalds/Colbourne/Karski/Williams (1991), *Nottingham Heritage Area Study*. Unpublished report.

7 Waagstraatcomplex and Hoofdstation, Groningen: Consequence or Cause of Place Identity?

M.J. KUIPERS and G.J. ASHWORTH

.... van wat hier door de jaren is verrezen
is veel weer door de jaren neergehaald
maar altijd werd door deze plek *het wezen*
van Gronings stad en ommeland bepaald
dat, steeds als men het nieuwe en het oude
opnieuw in de waagschaal legt,
*voor volgende geslachten blijft behouden
wanneer ook deze muren zijn geslecht*
(Den Hollander, 1998, p.63)

What has been erected here over the years,
has also, over the years, been demolished.
But this spot has always expressed
the spirit of Groningen city and region,
which, still as we weigh the old with the new,
will be maintained for future generations
whenever these structures have gone.
(Part of the poem written by J.P. Rawie,
translated by G.J. Ashworth, 2000)

Heritage and the Symbolic Restructuring of Places

Rawie's poem, written on the occasion of the official opening of the newly built *Waagstraatcomplex* in 1996, argues that the location in the city centre of Groningen, where the Waagstraatcomplex is now, has always housed the spirit of Groningen, city and surroundings. More notably it argues that this spirit also will remain for future generations, even when the current *Waagstraat* buildings have been long demolished. The sense of place is expressed through both a location and the premises constructed on it. This essence of the city is a product of the past but interpreted and projected by the structures and designs of the present. This raises numerous questions about the role and function of contemporary decision making and especially public sector local planning in creating or propagating, consciously or not, such senses of place. Place identities or *genius loci* means the feelings and meanings ascribed to a place by people, both as individuals and in groups. Here, the more objective characteristics of a place are brought together with the subjective aspects of the experiences that various actors have with that place. Kersten (1998) makes the obvious point that the quality that people ascribe to a historic building depends not only on its design or architectonic quality but also on the identity that people ascribe to it. According to Rose (1995:89) place identities are embedded in the social, cultural and economic circumstances that created them. The idea that this collective identity not only changes but also is changeable through deliberate action offers an interesting possibility and challenging responsibility to local planning authorities.

The case of the changing identities of the city centre in Groningen provides an opportunity for focussing upon two more specific aspects of these identities. First, selected aspects of the past play as heritage a critical element in place identity. Heritage planning focuses on the conscious use of physical and imaginary structures from our past in the present in order to help with shaping our future (Ashworth, 1994:1). In practice this could mean on the one hand the renovation and conservation of heritage buildings or on the other hand the adaptation and rethinking of the stories about or from the past in order to create a better built environment (Kersten, 1998). Secondly, such identities can be shaped, influenced or even concealed by deliberate planning action. How and why this is done, and the limits of its effectiveness, are the main questions posed here.

Urban Imagery

As long ago as 1960 Lynch examined urban imagery, by investigating the way townscapes are 'read' by urban populations. Most of all he wanted to decipher how particular physical aspects of the urban scene are perceived by different social groups. Simply said, he defined urban imagery as 'the perception of the city held by urban populations'. This was essentially descriptive but had implications for management through image creation and promotion especially to influence the readability of the cityscape. In particular he listed conditions for a desirable city structure, which should be a combination of clarity in its orientation points with surprise, variety and diversity. One logical step further is the recognition that the form of the city, its shapes, buildings and spaces, contains meaning as Rapoport (1982), and many others have argued in detail at length.

The forms of the city are therefore signs which communicate meanings through symbolism (Tuan, 1977). Places, as Eco (1986) has argued, possess attributes of both denotative functionality and connotative communicativeness. This has perhaps always been so and such a restatement does not in itself advance the argument. Such signs are what makes one place distinct from another, localities rather than points in an abstract spatial geometry and have thus always been consciously present, not least in what are now called 'signature buildings'.

The forms of the city and indeed the city as a whole may well convey meanings and it may be that 'the city is a discourse and this discourse is truly a language' (Barthes, 1986:92). The city may in this sense be a language (Barthes, 1970; Choay, 1970; Eco, 1972; Gottdiener & Lagoulopolis, 1986; Gottdiener, 1995; Ashworth, 1998). The initial problem however, the solution of which is a prerequisite of any effective intervention, is to determine which signs convey which meanings to which readers. Meanings are conveyed through codes and thus both encoding by producers and decoding by consumers are required. These code systems are neither universal in space nor stable over time. The producers of the city's forms may themselves be unclear, ambiguous and pluralist in their messages. In addition the physical signs used to convey such coded messages are generally quite deliberately designed to be robust enough to survive over long time periods. The same forms may be reinterpreted many times during their physical life span or their messages may become indecipherable, unintelligible, irrelevant, ambiguous in meaning, volatile or for various reasons unacceptably distasteful, and thus repellent to particular readers. There is also very little evidence that those who read the conserved city and its heritage actually receive the projected messages as intended either by the original encoders or by modern interpreters. Simply, 'one of the hallmarks of man-environment research is the realisation that designers and users are very different in their reactions to environments and their preferences, partly because their schemata vary' (Rapoport, 1982:15-6). The code books are likely to be different and the meanings of the encoders can only be explained through explanatory 'marking'.

Thus we arrive at a city whose conserved forms express many messages, intended and unintended, legible and illegible, which have been plurally encoded by socially pluralist societies and also are now decoded pluralistically. These in turn contribute to a place identity that is itself unclear, ambiguous, and continually changing. If communities create place identities and in turn depend upon such identification for their cohesion, then unsurprisingly diverse, pluralist communities create and reflect equally diverse and pluralist places. This 'diversity, variety and richness of popular and local discourses' (Featherstone, 1990:2) is unlikely to 'playback systemicity and order' so much as confusion and disorder. However the argument here depends on the discovery of just such order so that goal directed intervention could occur.

The idea of intervention is also not new. Boorstin (1961) claimed that imagery is influenced by what he terms 'pseudo events' which are presented through newspapers, films, television and advertisements which create a, 'thicket of unreality which stands between [Americans] and the facts of life' (Pons, 1975:4). This is a narrow view, which assumes the existence of single ascertainable reality from which promoted images diverge. It does however introduce the idea of image creation. The ideas of Strauss (1961) about urban imagery encompass more than those of Lynch and Boorstin together, namely, 'the whole range of established representations of the urban world through which people interpret their own particular cities and our Western urban habitat in general' (Pons, 1975:4). According to Pons (1975) Strauss focused on the way urban dwellers continuously use, create, develop and redefine images of their towns, as well as of the variety of identifiable social 'worlds' within them.

According to Strauss (1961) the city is not only passively perceived but also actively conceived. That is, we interpret our surroundings, we use them, we select aspects of them, and, implicitly or explicitly, we conceive or even construct images of them. Imagery consists of constructions in our minds or within our imaginations and memories. However images are not static but change continually, because of the changes in ourselves and our vantage points but also because they are produced and used in changing situations (Pons, 1975).

Lowenthal (1985) argues that the recent collective image of the past determines to a large extent the selection, renovation and presentation of the monumental past. The way preserved buildings look today is a consequence of how the restorers imagined them (Denslaken, 1994). In this way the renovation of a monument could be seen as a change that takes place on the basis of our image of the past. More widely, many plans for the built environment reflect the way that people look at it. In other words they tell us about the prevailing views with respect to the built environment.

Groningen: Towards the 'Bologna of the North'

The city of Groningen is a medium sized (c. 170,000 inhabitants), multifunctional city serving a dominantly rural hinterland of around 500,000 people. It has functioned for almost a millennium as the main government, education and service centre for the northern Netherlands with market, agricultural processing and industrial functions (Kooij & Pellenbarg, 1994). However in more recent years it suffered from deep seated chronic economic difficulties, caused in part by its relative isolation from the core areas of The Netherlands and Europe, the decline in the regional importance of agriculture and the stagnation of most of the regional industrial base (sugar beet, dairy and tobacco processing, ship building and repair, cardboard and paper manufacture). The economy of the northern region therefore has long been typified by low incomes, high unemployment, and high dependency on social security receipts relative to the rest of the country and by a perceived isolation, social conservatism and cultural traditionalism. The significance of this sketch of, a not particularly exceptional in Europe, economic and social condition is that it operated as the underlying imperative powering the developments in cultural productivity, physical planning and design and above all place image and identity that are described here.

As early as the middle 1970s it had become clear that the traditional activities and role of the City had been severely weakened, that new activities and directions for development had to be found and that the City and its region was fundamentally disadvantaged compared with other regions in The Netherlands and neighbouring countries. The compensating strengths for the economic weakness and social backwardness described above, were a strong sense of distinctive local identity and a cultural, educational and even artistic and monumental legacy inherited from a thousand years of exercising what amounted to uncontested capital hegemony in the North. The decision was therefore made, not at any one traceable moment nor by any specific local government agency or individual, but by an emerging consensus that the City's future lay in the reshaping of the city physically, economically and perceptually. The existence of regional cultural facilities (symphony orchestra, city concert hall and art complex, theatres, art and music academies and the like) as a result of its regional capital status and of major higher educational institutions, crowned by the national university (the second oldest in the country) provided a notably young and highly educated population as both market and product. The intention was that on this base would be constructed a new city focussed upon culture, the arts, entertainment, education and science of national and even international allure.

Of particular interest here is that urban design in general and heritage in particular were to play catalytic and promotional roles. Heritage here should be understood to include not just preserved monumental buildings and conserved areas but a conscious use of past associations and references. The large number of nationally designated monuments has indeed almost doubled in the past decade (see Figure 7.1) and the entire central area of the city is a conservation area. However there has been no question of shaping a fossilised heritage city excluding modern functions or even significant new developments. Indeed the past has been deliberate invoked and even recreated in the service of the present and in pursuit of the futures outlined above. A new overall design standard for city centre public space introduced new materials (the now famous 'yellow brick roads' of Groningen), street furniture, public artworks and non-motorised routing systems that were expressly intended to invoke a past and strengthen a local distinctiveness through an invented vernacularisation of the streetscape. The two main policy initiatives of the local authority, which are the concern of this chapter, were first, the creation of a number of 'flagship' projects and secondly the production and implementation of a new comprehensive plan for the inner city. The first included major new buildings, such as the *Gasunie*, public library and especially new museum, all of which were visually dramatic, architecturally experimental but which in various ways rejected functional modernism in favour of an eclectic evocation of selected aspects of the past whether imagined or not. In addition the rediscovery of the past included a city boundary marking project and the recreation of the late nineteenth century railway station. All of these were 'signature' projects in that they clearly signalled the existence of a new approach and new elan that combined a recognition of the continuities from the past with an experimental self-confidence in a future urban renaissance that was distinctly Groningen. The plan for the management of the form and functioning of the central city, known as *Binnenstad Beter* (Gemeente Groningen, 1992), was begun in 1993 and its implementation is still in progress.

Both the initiatives for remarkable signature building and for new city centre management are combined most notably in the Waagstraat project, sloganised by the City Council as 'a new heart for Groningen', to which we now turn our attention.

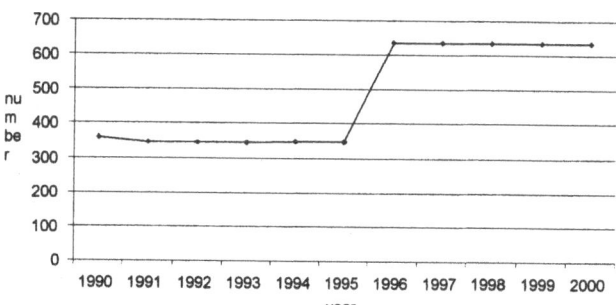

Figure 7.1 The number of officially designated national monuments in the municipality of Groningen (1990-2000)
Source: M.J. Kuipers

Waagstraat

The focus of the centre of the city of Groningen is the market square, the *Grote Markt*, which has operated as 'the living room of the city' for a thousand years. At its western end it is joined by the main traditional routeway into the City along the dry sand ridge, called the *Hondsrug*, that extends 50 km southwards. This *Hereweg*, still the main retailing street, was extended along the west side of the Grote Markt in a short commercial street, the Waagstraat. The 1945 three day battle of Groningen (Ashworth, 1995) was a divisional scale infantry and armour engagement focussed on a restricted area of the city centre which resulted in the total demolition of the west, east and north sides of the Grote Markt. Post-war reconstruction of the western side ignored the old street pattern and used the site for the construction of a new office block in 1962 (Den Hollander, 1998:61), in minimalist international style, as an extension of the existing town hall. Old and new were physically and functionally joined.

The decision to demolish the post-war building, restructure the space and rebuild in a different style and for different functions was prompted not by any physical inadequacy or decay of the existing buildings but in order to conform to the new self-image of the city as described above. Essentially the City wished to project to itself and to outsiders a different image than that created after the war.

The Results of Public Consultation

The population of Groningen was presented with four submitted designs for the Waagstraat location and the associated planning proposals for the entire public space of the Grote Markt. These were the plans of the offices of the architects Jo Coenen, Gunnar Daan, Adolfo Natalini and Coen van Velsen. All of the four plans presented, except that of Natalini, could be categorised as 'modernist', consisting of building blocks at various heights surrounded by geometrically sharp spaces. The Natalini-plan however could be viewed as 'traditional' in its choice of red brick, its height at 4 or 5 storey similar to the neighbouring buildings and inner city in general and its creation of smaller, intimate public spaces. Its detailing was strongly reminiscent of Italian renaissance styles, in its fenestration, projected rounded rooflines, and colonnaded ground floor. Its stylistic eclecticism in time and place led to its designation as 'post-modern'. The result was remarkably clear (Figure 7.2), 83.6% preferred the Natalini-plan: a small group of respondents liked the Daan-plan while the plans of Coenen and Van Velsen formed the least preferred options.

Among those favouring the Natalini-plan the central prevailing argument was that the plan 'fitted'. According to the *Projectgroep Waagstraat* (1991) these supporters had the opinion that the plan formed a natural link with the inner city; with the overall street pattern, with the existing retained historic buildings such as the *Goudkantoor* (Figure 7.3), the old city hall, and the *Martinitoren*, as well as the new library in the *Oude Boteringestraat* built by Grassi and with the south side of the Grote Markt. Respondents who supported the Natalini-plan frequently used the following comments. 'The plan has atmosphere, cosiness, warmth.'; 'The plan is less time-bounded.'; 'In a positive way, it is a little traditional.'; 'The reference back to the past in a contemporary style is attractive.'; 'The design gives the city centre its heart back.'; 'The design is not world-shocking, but pleasant and intimate.'; 'It approaches the idea of a living room for all inhabitants of Groningen'. There were of course also negative comments but these were far fewer and in any event focussed on much the same features of traditionalism, unadventurousness and even kitsch.

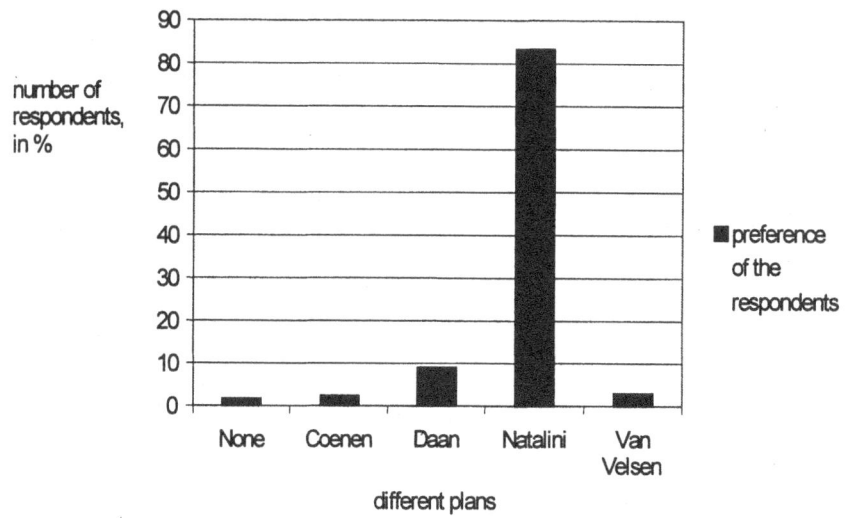

Figure 7.2 Preference of the respondents
Source: M.J. Kuipers

Figure 7.3 Adolfo Natalini's Waagtstraatcomplex with the integrated historic Goudkantoor
Picture: G.J.R. ten Kate (March 2001)

The local government of the City also empanelled a Waagstraat Commission of Judgement, composed of a number of independent experts in town planning and architecture, to help with the making of a choice between the four plans. Its brief was to contribute to the discussion about the programme as a whole, consider the relation with the Grote Markt as an entity and judge the quality of the architecture itself. It is remarkable that ultimately the commission chose the plan of Gunnar Daan and not the plan of Adolfo Natalini thus directly contradicting the popular view. It had the following opinion about the Natalini design:

> The architecture is disappointing. In this design the identity of the city has been put on a level with nostalgia: a false sentiment that is misplaced on this location. Here the architect tries to create a design, which has its roots in the local tradition. The accurate presentation recalls the suggestion of an authentic historical city scene, however the facts are otherwise: both building-typology, façade-architecture and street scene miss every historical character and offer a few points of contact for the intended urban identity.
> (De Beoordelingscommissie Waagstraat, 1991, p. 5)

Clearly the 'experts', unlike the citizens, had little time for more traditional styles and materials, finding nostalgia to be a 'false sentiment' while simultaneously believe in the existence of an appropriately 'authentic' reference in time and space.

The Groninger Hoofdstation: Rebuilding and Return

The Influence of the Economic Prosperity in the Late Nineteenth Century

The period in which the station was built (1893-1896) was a prosperous one. It was the time of the industrial development, especially the potato flour industry, which was founded by W.A. Scholten. The city theatre was being built and people wanted also a station that had style and gave a reflection of the economic prosperity and modernity (Salverda, 1998). The reason that the 'Dutch Railways' paid much attention to the design and building of railway stations, both in Groningen and in other cities, is that the train as a means of transport had a specific image for the average citizen. In that time travelling by water was more popular, so the design of the stations had to contribute to the image improvement of the train as a means of transport.

The Influence of the Functional Sixties

In the 1960s the opinion about former construction and buildings changed. In that time the reconstruction in The Netherlands accelerated. According to Dijkstra & Akkerman (1999:71) the 1960s were marked by functionality, housing shortage and new housing estates. So buildings had to be useful, functional, modern and sober. Many buildings were demolished, reconstructed or adapted as the economy and society modernised. 'If it only looked nice and authentic from the outside, you could do anything with the inside of the building' (Dijkstra & Akkerman, 1999). On the basis of this thought many buildings underwent a big change in The Netherlands, while the exterior was kept intact.

Although there was already a general view that the station building had great architectural value, it applied only for its exterior. So, in 1968 a drastic rebuilding of the interior of the station building took place. There are three main reasons for this: the 'spirit of the age' ascribed above, various rational reasons in relation to the functionality of the building and the bad condition of the ceiling. A new lowered ceiling was built, so that the beautiful artefacts were covered completely. The old walls were covered with board-material. Because of the coming of modern ticket-offices, a shop, a bank office, a travel bureau and a florist's shop in the hall, there was only one third of the floor space left. The side wings became inaccessible from the hall.

The Influence of the Late Twentieth Century

In 1994 Rob Staal took the initiative and developed plans for a restoration of the hall of the *Groninger Hoofdstation* (Figure 7.4). In 1996 a restoration fund was established, existing of both private and public funding. According to Dijkstra & Akkerman (1999:57), the recent interest in buildings like the Groninger Hoofdstation is influenced by the revival of the valuation of art from the nineteenth century that has taken place in The Netherlands lately. Salverda (1998) argues further that it is not very accidental that the restoration begun in 1998. Twice earlier it was tried in vain. The reason why it did not fail this time was probably because of the fact that two different developments came together in this project (Salverda, 1998). At the same time the Dutch Railways wanted to adapt the station to the new conduct of business and there was a group of people that had the idea of restoring the station hall. Together they created the moment of realisation (Salverda, 1998).

The Groninger Hoofdstation in its Surrounding Area

The surrounding area of the Groninger Hoofdstation, also called 'the station area', has changed constantly in the past century. The buildings around the station changed in both form and function. In the beginning, the station building was surrounded by a few dwelling houses, which were subsequently replaced by office buildings.

According to Dijkstra & Akkerman (1999) the Groninger Hoofdstation remained vital to the townscape and played a major role in the planning discussion during the different processes of change. Questions arose about what did fit and what did not fit near such a massive monumental building and about which of the existing structures near the station could or could not be sacrificed for new building. According to Dijkstra & Akkerman (1999) it was the coming of the Post Office to Groningen that made it necessary for the Groningen City Council to look at the station area in a critical way. The Post Office was only prepared to move from the Randstad to Groningen, in accordance with central government relocation policy, if they got a new building near the station (Figure 7.4).

Figure 7.4 The Groninger Hoofdstation and a part of the Post Office building seen from the Werkmanbrug
Picture: G.J.R. ten Kate (March 2001)

At the same time on the national level there was a trend of stimulating employment near stations, in order to ease traffic problems. Although the population of Groningen always has been supportive of an image of a large town, it was opposed to the demolition of some buildings near the station, which were perceived to be vital to the townscape (Dijkstra & Akkerman, 1999). Many of the inhabitants of Groningen were also of the opinion that the new substantial office buildings dominated the smaller station building too much.

Opposite the station is the widest part of the *Verbindingskanaal* (linking canal) known as the *zwaaikom* (turning point). In the 1980s the future of the zwaaikom has been hotly debated. Some wanted it to be left as it was and others wanted it to be built over. The final plan started when the municipality of Groningen received a gift from the Gasunie to build a new museum. As a result, in 1994 a new and unique *Groninger Museum* (Figure 7.4), designed by Mendini, was opened in the zwaaikom itself.

The relation between the Groninger Hoofdstation and the Groninger Museum is remarkable in two respects. On the one hand there is a symbolic link between them. As a consequence of the new bridge, the *Werkmanbrug*, (Figure 7.4) over the canal, in Mendini's design of the new Groninger Museum, the station hall received a symbolic meaning of 'entrance' to the inner city. In turn the Werkmanbrug, together with the

Folkingestraat (see Chapter 17 in this volume), could be seen as a stately corridor to the inner city. The bridge has had a fundamental influence on the recent walking routes from, to and through the city (Dijkstra & Akkerman, 1999:89).

On the other hand however there is a contrast between the two buildings. This contrast lies mainly in the difference between their architectural meanings. The Groninger Hoofdstation expresses a nostalgic romanticism. It reminds us of, or even returns us to, the time the station was built. Its location and its design are not arbitrary. They are well considered on the basis of demands for efficiency, functionality and accessibility. However, the demand for beauty also played a role in the design. The Groninger Museum can be seen as a post-modern phenomenon. Both its design and its location are not logical; they are both unexpected. In line with this, the Folkingestraat also expresses a sense of post-modernism through its arbitrary mixture of different sorts of unexpected shops, such as a shop with percussion instruments, a shop with Turkish food, a shop with Asian articles. Thus the renovated station, the new museum and the burgeoning 'art and culture' shopping of the Folkingestraat together comprise the new linear entry to the new Groningen.

Conclusion

Three final points need to be made. First the harmonious relationship of old and new, renovation and rebuilding, site and structure was never automatic or self-evident. It would have been all too easy for the mix to produce a disjointed disharmony. Secondly, the local government played a decisive role in each of these developments. There were of course many other actors and interest groups involved who contributed finance, structures and ideas but it was the local city government who initiated, co-ordinated and stimulated change. Finally, buildings and designs are ultimately valued by those who use them and their identity in the last instance will depend upon the meanings that users attach to them over time. The planning slogan was 'building a new heart for Groningen': the success will be determined by the beat of the lifeblood of people through it.

References

Ashworth, G.J. (1994), *De pijl des tijds in het ruimtelijk doel*, Geopers, Groningen.

Ashworth, G.J. (1995), 'The city as battlefield: the liberation of Groningen', April 1945, *Groningen Studies*, vol. 61.
Ashworth, G.J. (1998), 'The conserved European city: the meaning of the text', in B. Graham (eds) *Modern Europe: power, culture and identity*, Arnold, London.
Barthes, R. (1970), 'Semiologie et urbanisme', *Architecture d'Aujourdhui*, vol. 42, pp.11-13.
Barthes, R. (1986), 'Semilology and the urban', in M. Gottdiener, A.P. Lagopoulis (eds), *The City and the Sign: an introduction to urban semiotics*, Columbia University Press, New York.
Beoordelingscommissie Waagstraat, de (1991), *Rapport Beoordelingscommissie Waagstraat*, Gemeente Groningen, Groningen.
Boorstin, D. (1961), *The Image*, Atheneum, New York.
Choay, F. (1970), 'Remargues a-propos de semiologie urbaine', *Architecture d'Aujourdhui*, vol. 42, pp. 9-10.
Denslaken, W. (1994), *Architectural Restoration in Western Europe: controversy and continuity*, A&NP, Amsterdam.
Dijkstra, J. and Akkerman, H. (1999), *Groningen Hoofdstation Centraal*, Regio-Projekt Uitgevers, Groningen.
Eco, U. (1972), 'A componential analysis of the architectural sign', *Semiotica*, vol. 24, pp. 97-117.
Eco, U. (1986), 'Function and sign: semiotics of architecture', in M. Gottdiener and A.P. Lagopoulis (eds), *The City and the Sign: an introduction to urban semiotics*, Columbia University Press, New York.
Featherstone, M. (1990), *Global Culture: nationalism, globalisation, identity*, Sage, London.
Gemeente Groningen (1992), *Binnenstad Beter*, Dienst RO/EZ Groningen.
Gottdiener, M. (1995), *Postmodern Semiotics: material culture and the forms of postmodern life*, Routledge, London.
Gottdiener M. and Lagopoulis, A.P. (eds) (1986), *The City and the Sign: an introduction to urban semiotics*, Columbia University Press, New York.
Hollander, F. den (1998), *Stadswandelgids Groningen*, Uitgeverij De Geus bv, Groningen/Breda.
Kersten, R. (1998), 'Beelden van verleden', *Rooilijn*, vol. 9, pp. 453-458.
Kooij, P. and Pellenbarg, P. (eds) (1994) *Regional Capitals: past, present, prospects: Ghent, Groningen, Muenster, Norwich, Odense, Rennes*, Van Gorcum, Assen.
Lowenthal, D. (1985), *The Past is a Foreign Country*, Cambridge University Press, Cambridge.
Lynch, K. (1960), *The Image of the City*, The M.I.T. Press, Cambridge, Mass.
NCM (1990-2000), *NCM Monumentenjaarboek*, Amsterdam, Stichting Nationaal Contact Monumenten.
Pons, V. (1975), *Imagery and Symbolism in Urban Society*, University of Hull, Hull.
Projectgroep Waagstraat (1991), *Resultaten publieksenquête Waagstraat*, Gemeente Groningen, Groningen.
Rapoport, A. (1982), *The Meaning of the Built Environment: a non-verbal communication approach*, Sage, Beverley Hills.
Rose, G. (1995), 'Place and Identity: a sense of place', in D. Massey and P. Jess (eds), *A Place in the World?* Oxford University Press, New York.
Salverda, E. (1998), 'Stationshal Groningen in oude glorie hersteld', *Broerstraat 5* vol. 1, pp. 6-7.
Strauss, A. (1961), *Images of the American City*, The Free Press, New York.
Tuan, Y.F. (1977), *Space and Place: the perspective of experience*, University of Minnesota Press, London.

Theme 2: The Heritage Site as Attraction

ANGELA PHELPS

When looking at extraordinary buildings of the types described in the following three case studies, it is not always appreciated that their survival is the result of a precarious balance of expediency, opportunism and neglect. Historic buildings may be of interest for their architectural merits, their association with people and events, or as beautiful constructions to be enjoyed for their aesthetic qualities. The balance of interests may change over time, and consequently conservation decisions are unlikely to remain unchallenged. As the examples in this section demonstrate, conservation of substantial properties can only be effective when there is a consensus regarding their importance, and for the largest properties, this requires acceptance as part of a national heritage.

While properties remain in private ownership, their retention is largely a matter of continued maintenance and functional re-modelling. Indeed it is only in recent decades that the idea of restoring an historic building to the appearance of its original construction has been seriously promoted. In previous centuries owners had little compunction about demolishing outdated buildings and replacing them with something more fashionable. All the examples in this section raise questions concerning the processes of conservation: should early nineteenth century restorations at Wollaton Hall be removed; is the twentieth century construction at Gunnebo to be considered authentic because it utilised original plans and traditional methods; should the surviving walls at Naarden be considered of greater value than the reconstruction at Bourtange?

The purpose of listing is to identify properties worthy of more focused conservation and protection from demolition. However, there is a growing opinion that the operation of such systems, with their emphasis on structural conservation, is too restrictive. A forward looking 'renewal' is now advocated finding multiple function re-use that will not simply preserve the buildings, but bring back energetic use. To be successful, this requires a less dogmatic approach to listing that recognises social and economic context as well as architectural merit, and prioritises local over national interests.

Historic properties that have survived into the modern period will already have faced successive adaptive re-use. Even where the use has remained the same, there is still likely to have been phases of renovation. The prioritising of one period for the purposes of heritage reconstruction will inevitably result in the obliteration of any trace of the site's subsequent history. Where form and function are closely related, changing the function may seriously under-mine understanding of the surviving forms. This is a particular problem where the context of the site has been lost, as clearly demonstrated in the case of the fortress towns of The Netherlands. Presentation as anything other than a fortress is not feasible, but the tourist appeal of such sites is limited. Nevertheless, once the original function of an historic property is relinquished, an alternative use must be found if the buildings are to be maintained in good order.

Whether in public ownership, private ownership or the guardianship of a conservation organisation, a common theme runs through accounts of current management. Historic properties are very expensive to maintain, and in every case the most pressing challenge is to find ways of generating income to pay for the upkeep of the buildings. The great country houses have always been a substantial drain on the resources of their owners. In the past their upkeep was part of the general expense of an estate, but as costs rise private owners retreat. Where properties have passed into public ownership, funding is still the key issue that under-pins, and under-mines, all other considerations.

Solutions resting on income from tourist visitors may provide additional pressures. Any property open to the public requires attention to visitor facilities, the minimum being car parking, toilets and security; income generation prompts the addition of shops and catering. Successive changes in the law have required attention to health and safety, food hygiene and most recently, disabled access, all of which may result in costly adaptation. It is almost inevitable that the needs of public access will create conflicts with conservation. Restrictions may seriously impair the potential of developing a visitor attraction within the property, which is why the most successful ventures have substantial other attractions: theme parks, galleries, working farms, events and temporary exhibitions. These activities divert investment and their commercial nature may compromise the very purpose of conservation.

8 Developing an Historic Monument: Reinventing the *Villa Rustica* of Gunnebo

INGER ERNSTSSON and BENGT O.H. JOHANSSON

Figure 8.1 Villa Gunnebo with reconstructed garden
Picture: Inger Ernstsson 2000

This chapter investigates a quite typical example of present trends within the field of heritage management. The case concerns a nationally well-known historical building outside Göteborg, and its development into an attractive heritage site. The notion of development as understood in this case included not only the physical conservation of this important and spectacular building, but also the promotion to the citizens of its cultural and historical values; to offer vocational training to craftsmen, to further integrate the surroundings of the building into the management of the site; as well as to attract as many visitors as possible. The threefold objective was to raise public awareness and interest while simultaneously offering recreation and amusement; to motivate the municipal owner to run the site in a fashion proper to its cultural values; and to raise funds for the continued maintenance and conservation of the site, without compromising too much its perceived authenticity. Such a multifaceted agenda could not but create dilemmas, and one such was played out when a new service wing was built as a meticulous reconstruction of a never built project, using authentic materials. This brand new wing was then adapted to modern uses under a heritage control similar to those ideally applied in conservation. In this case, we will argue that the solution chosen challenges the concept of authenticity in a way never anticipated in existing preservation charters (for example the Venice Charter of the International Committee of Monuments and Sites 1964, The European Charter of the Architectural Heritage of the Council of Europe 1975 or UNESCO's celebrated World Heritage Convention, 1972).

Background

In 1778 John Hall Sr., a well-to-do merchant of English extraction bought the farm Gunnebo not too far from his hometown Göteborg (Baeckström 1977). Hall was enjoying the local economic boom of the late eighteenth century, making his fortune from trading iron, timber and fish oil. He had married the daughter of another prominent merchant who was engaged in the Far East market as superintendent of the Swedish East India Company. The Hall couple played an important role in the social life of the Göteborg bourgeoisie, and their house in the city belonged to the more sumptuous buildings. A summerhouse in the countryside was however at this time beginning to be a *sine qua non* for the emerging new class of burgers. The immediate surroundings of Göteborg with their naked rocks and uninviting heaths were not particularly suitable for such a project to the late eighteenth century eye. In contrast, the Gunnebo farm, situated a comfortable 10 kilometres from the city in an attractive lake district, provided the right kind of pastoral landscape with a mixture of fields, meadows and arable land.

No architect could be better suited to design Mr. Hall's summer house than the city architect of Göteborg, Carl Wilhelm Carlberg, who at the time had just returned from a five year study tour in Europe. Carlberg's tour had included not only Italy and France, but also England, and this most probably made Hall find him particularly well-suited for the project. They had also worked together in one of Hall's industrial projects. Together they created a summer residence of the highest quality, in many ways unique to Sweden, due not the least to its affinity with English Palladianism. Carlberg designed all new buildings on the site; the park and the garden; the interior decorations and the furniture, as well as the tiled stoves, typical for Sweden. The project resulted in a host of drawings, which miraculously have survived (Baeckström 1977). Considering the comprehensive character and quality of Carlberg's work, there is no wonder that Gunnebo was already highly admired by Hall's contemporaries.

Unfortunately John Hall's son, John Hall Jr., was totally uninterested in business life, and consequently the trading house collapsed in 1807. Twenty-five years of serious decay followed, and the orangery, the wings and the hermitage burnt down, the park was used for grazing, and leaking roofs of the main building added to the misery. However luckily enough, the owners taking over from John Hall Jr. were interested in the site and the unique architectural and cultural values created by the first owner and his architect. The last private owners in particular devoted a large part of their lives to an attempted re-creation of the lost paradise.

Gunnebo in Municipal Hands

The Gunnebo estate is located in Mölndal, a municipality adjacent to Göteborg, which in 1949 took over the responsibility for Gunnebo from the then deceased Baroness Sparre. The goal – expressed in a letter of intent – was to preserve for the future the extraordinary high cultural values manifested in the estate of Gunnebo. The municipality agreed to have the building listed in a form equivalent to the English Grade 1*. There is no doubt that the primary municipal goal was to preserve the estate and make it publicly accessible. The municipality installed a competent board to govern operations, and during the 1950s, ambitious and highly qualified restorations of the main building and its interiors followed. The restoration and reconstruction work was preceded by strict building documentation and was facilitated by the use of old inventories and original drawings. However when, a few years later, it came to the reconstruction of the large exterior stair towards the garden, the established practice of using authentic materials was dropped and a construction in concrete was chosen – although garnished with stone balusters. The last major undertaking was when the park was partly reconstructed by Walter Bauer, by then the most renowned Swedish landscape architect and restorer of parks and gardens. Bauer strictly followed Carlberg's surviving designs.

Through industrialisation, the municipality of Mölndal became a predominantly working class community, and it is of interest that the municipalisation of the Gunnebo estate was manifested by its use as the destination for the traditional First of May parades organised by the ruling Social Democratic party. The parades had Gunnebo as their goal throughout the 1950s and up to the early 1960s, and thus, in a somewhat ironic fashion, working class people once a year expressed themselves as the owners of the former capitalist estate. With regard to the interior spaces, an important but not unproblematic function was the use of the building for municipal representations. Public access was accommodated through guided tours, concerts and other cultural programmes, and a café was installed in the basement of the main building. Also, the park became a popular recreation ground open to the public and was, as such, appropriated by the citizens.

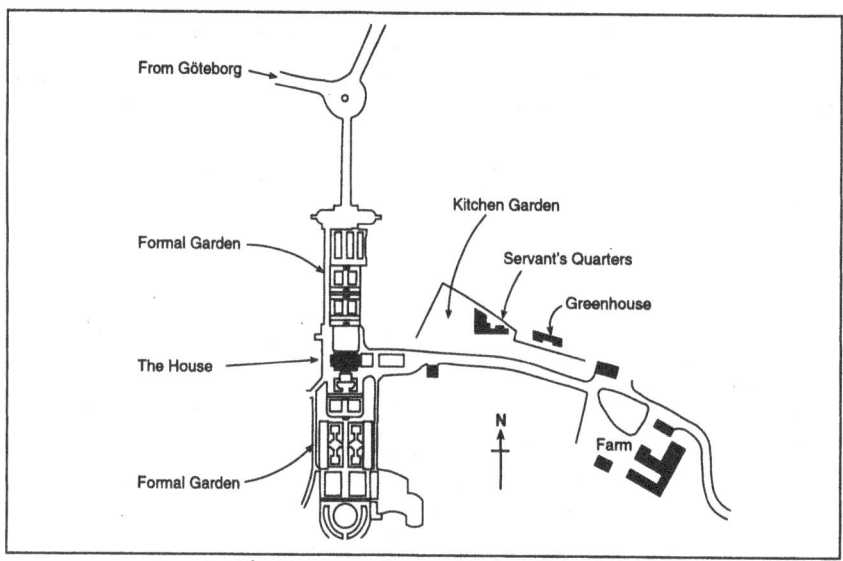

Figure 8.2 Plan of Gunnebo reconstructed
Source: Gunnebo House and Gardens

In the early 1970s, the municipality decided to split up the management of Gunnebo between different administrative bodies in order, one would assume, to rationalise up-keep and save money under the assumption that the highest competence in technical matters was to be found inside specialised municipal departments. The park administration was thus entrusted with the park, the housing department with the buildings, and so on. During this period, and despite the fact that the most necessary repair work was made every now and then, Gunnebo again gradually decayed (Jonsson & Tollbom 1981), while simultaneously visitors became fewer in number.

Gunnebo – Back to the Eighteenth Century

In 1995, Mölndal municipality realised that something had to be done to revitalise Gunnebo. In this situation, a chain of events was triggered off when the state – in response to a demand from Gunnebo for economic support for the restoration of a few sculptures in the park – requested a comprehensive plan for reviving the park and garden. The local museum was engaged in preparing visions for the future of Gunnebo. However, the immediate goal was to engage a project leader on a two years contract with the immediate task of preparing a business and management plan.

Key notions in the plan were to make the public more interested in Gunnebo, to use it as a source for disseminating historical knowledge, and to attract a larger share of the growing field of cultural tourism. Following upon this plan a more detailed action plan was decided upon. This plan identified five different areas to be developed based upon the overall objective that had been defined already in the 1940s, namely to preserve the extraordinary high cultural values embedded in the estate. The main building, the park and garden, as well as the adjacent farm should be developed to offer visitors' recreation and inspiration, as well as new experiences and knowledge. This meant that Gunnebo should be revived and be made to live a life as natural as possible, which in its turn meant that long lost functions in the life of the eighteenth century estate should be evoked. To this one important goal was added: Gunnebo should serve as an educational enterprise not only for visitors but also for all craftsmen involved in the project.

The five areas of the action plan were defined in relation to the following observations:

- *The main building* was found to be in need of further restorations as well as regular maintenance, but also further measures to avoid wear due to an increase in number of visitors.
- *The formal garden* was to be reconstructed to its original eighteenth century character in order to enhance the site, both with regard to spatial planning and plants used.
- *The kitchen garden* and *the greenhouse* were to be reconstructed as prerequisites for future reconstruction work and for accommodating a gardener's workshop.
- *The servants' building* was to be reconstructed in order to make it possible to move this service out of the main building and for the café and an additional restaurant to offer a wider selection of food and drinks, mainly based on products from Gunnebo itself.

- *The cowhouse* was in need of renovation in order to create space for a small stock of typical domesticated animals.

A further goal, typical of the 1990s, was added: the development of Gunnebo as a site should respect the principle of recycling and make use of natural materials. This goal is essentially, of course, in line with how a household was run in the eighteenth century and thus adds another dimension to the educational effects of the whole revival project. In a simple way, the concordance between modern principle of recycling and the ways in which sparse resources were employed in an earlier historical period offered a great pedagogical opportunity in the context of presentation.

The financing of the project exemplifies another feature typical of the 1990s with its blending of money from the Swedish state, the municipality and the European Community. The action plan was by necessity given a sequential logic in which one undertaking was dependent upon the prior finalisation of the preceding one. Thus, (1) the park could not be maintained without a functioning garden service, and, (2) this service must have a kitchen garden at its disposal. Further, the principle of recycling could not be respected unless there was manure at hand for natural fertilising of the kitchen garden. From this follows (3) that live stock had to be purchased and their upkeep provided. Leftovers from the kitchen garden and the restaurant could be fed to the animals or be composted, in both cases adding to the natural recycling of materials. Products from the garden were to be used in the restaurant. The reciprocal dependence between the servants' (service) building and the restaurant is obvious. Below, we will see how this simple logic inevitably led the project into decisions that were partly contrary to the strong preservationist position taken in the beginning by the project management. Even more decisive with regard to such deviations was, as we shall also see, the way in which the project was financed.

Another underlying assumption that was never challenged was active in the choice of period to be reconstructed. The reputation of the house had always been associated with its origin. Changes made during the lifespan of the estate were never deemed to be of historic significance. Without having been articulated as such, this choice was made already at the time of the first restorations in the 1950s, but could, of course, be questioned. Particularly from the point of view of conservation, the impact of the visions and ambitions (or lack of such) of the sequence of owners might have been of interest to preserve. There is on the other hand no ground for arguing that later periods were more interesting than the time of origin. The result of the concerted efforts of John Hall Sr. and Carl Wilhelm Carlberg did indeed impress people from the very beginning, and soon made Gunnebo valued as one of the more architecturally interesting buildings from the late eighteenth century in Sweden.

Financing the Campaign

In 1995 Sweden was a recent member of the European Union. The membership offered new possibilities also within the conservation sector. For example, the structural funds, Goal 3, opened up for partnership between EU and a member state in a way that suited the intentions of the Gunnebo action plan. A prerequisite for access to this EU fund was that the project had to result in both higher employment and higher competence of the work force through educational effects. The Gunnebo project was granted approximately 45 million SEK for the period 1996 to 1999 (EU 22 million, the Swedish state 9 million, and the municipality 14 million). The funding had to be used during this short time span, since one condition for Goal 3 grants was that a project should help to diminish acute unemployment among construction workers and at the same time help to increase their competence in a developing market for restoration activities. This condition created serious problems in the realisation of the project. From having been designed as a goal-driven process with a certain sequential logic, the actual process turned into a time-driven exercise defined by administrative stipulations. In this situation, it was not possible to simultaneously fully live up to the high standards set with regard to quality in education and execution in the learning process, and finalise as much as possible of the project within the allowed time-span. In the long run and when the severe unemployment in the building sector gradually eased and more jobs were offered on the regular labour market, it also turned out to be difficult to find motivated and talented craftsmen for the demanding project.

The Dilemma of Reconstructions – the Servants' Building

The purpose of the new servants' building was to offer better service to the public by expanding the café, adding a restaurant, and make it possible to offer a wider selection of food and drinks based on what was produced in the kitchen garden and by the farm. Expanding the coffee shop in the basement of the main building was out of the question since it would have seriously damaged the heritage values of that building. Carlberg's original design for the kitchen garden area showed a servants' building, which also defined the surrounding space in an architecturally satisfying way. It seemed obvious that the new building should build upon this original spatial concept. Since Carlberg's original drawings were accessible, the question emerged 'why not also let the original architectural design guide the solution?' However, in an orthodox interpretation, the famous Venice Charter on the restoration of architectural monuments would rather have suggested a modern building erected in modern materials and techniques in order not to make the public confuse the old fabric with new additions. However such a building would not have fitted with the ideas expressed in the action plan regarding the recreation of the ecology of the Hall family's household. The daring decision was made to use Carlberg's original design. In order to show not only a replica in the shape of the once proposed building, the new building – an imaginary reconstruction – was to be executed in 'authentic' materials and with 'authentic' building techniques. The notion of authenticity was here related both to a non-site specific knowledge gained from studies of existing buildings of the period, as well as to the extant main building, which was conceived of as a key source of site specific knowledge that also allowed access to solutions particular to Carlberg's professional practice. The craftsmen could consult Carlberg's extant design on site with regard to everything from the qualities of the paint to profiles of the carpentry.

  The ambition to erect new buildings at Gunnebo in accordance with eighteenth century principles included such techniques as making natural stone foundations without mortar, constructing wood panelled timber walls, and employing traditional roof tiles combined with roof details in iron sheeting. Timber, for example, was cut with axes on site. When sawn timber was used, such as in floors and ceilings, it was trimmed with hand planes in order to create the same appearance as if it had been done in the eighteenth century. In this way the craftsmen were trained in the use of traditional materials as well as in working with traditional tools. This approach meant that the construction of the service building could be used for the kinds of educational purposes that were an explicit condition for receiving grants from the structural fund of the EU. The construction workers, carpenters, painters, and other involved professions, were through their participation in the project to receive thorough training in traditional building techniques much needed in future restoration activities.

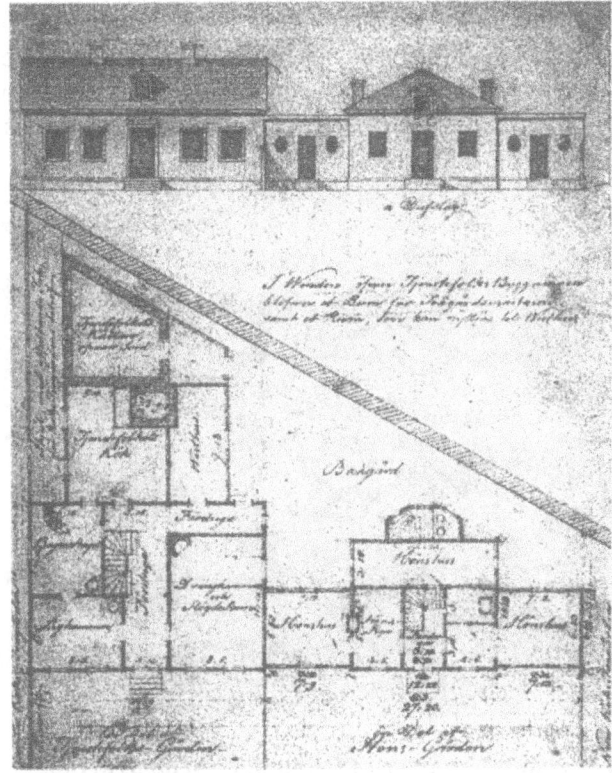

Figure 8.3 Servants' quarter at Gunnebo
Source: Original design by Carlberg

The service building was to be given different functions compared to Carlberg's original project. In this situation, the decision made was to treat the new building as it would have been treated had it been an authentic eighteenth century construction. That is, the 'imaginary reconstruction' was here treated as a 'new original', and present demands on functions and utilities to be fitted into this 'new original' were to be accommodated without damaging its imagined cultural values. The modern restaurant kitchen was carefully fitted into the kitchen area of Carlberg's imaginary building around the huge 'reconstructed' stove with its oven for baking bread. Thus, changes and additions to the 'new original' building were done in such a way that they should be clearly experienced as modern additions but also serve as a practical demonstration of important principles in conservation to the benefit of the craftsmen involved (not to mention that those principles also give flexibility to all kinds of buildings). Furthermore, all such changes and additions should be reversible and give the same impression as new installations in old buildings. The principles and procedures were the same as in a professional restoration of an existing building. On several occasions, this approach meant that exemptions from ordinary building codes were needed, and usually also given after some negotiations with the authorities.

Figure 8.4 Servants' quarter at Gunnebo, reconstruction
Picture: Inger Ernstsson 2000

With regard to other reconstruction and restoration work at Gunnebo, it was decided that Carlberg's own drawings were to be followed, but that these should be checked by thorough on-site investigations in order to assess whether his designs had really been followed in the eighteenth century execution of his project. Archaeological excavations were thus carried out in the park and in the kitchen area before reconstruction work started. The excavations in the kitchen area showed that the service building Carlberg had designed had never been built. Instead, the foundation of another building was found close to the location selected by Carlberg. In all probability, rather than engaging in the erection of a new building, the old farm house had been retained by Hall and turned into a servants' building. And in fact, a sketch by John Hall JR depicted that very farm house.

When the excavations were made, the project had already reached an advanced stage where the financing (and to some extent the whole idea of the Gunnebo revitalisation) had come to depend upon the reconstruction and the possibilities that such a project offered with regard to financial support from the European Commission. This was a major reason why the decision was taken to go ahead and engage in the 'imaginary reconstruction' of the servants' building. Hence, a choice that could be called pragmatic was made, and one with many implications. As touched upon above, the knowledge on which the reconstruction work was based was not, and could not be, complete. The 'imaginary reconstruction' has therefore resulted in a building dependent upon a mix of general knowledge of the period and very particular site-specific knowledge. If the relevance of the former is problematic, the relevance of the latter is no less questionable since there is no guarantee that Carlberg would have employed the same design solutions in a building quite different from the extant main building in function, type and symbolic importance. In order to avoid this last problem details were chosen from less 'important' areas of the building like the attic and the basement. Technical solutions were regarded as neutral.

Accordingly, what we have is an imagined eighteenth century servants' wing constructed as it might have been constructed, modernised in an imagined restoration process; presented as a simultaneously authentic and non-authentic building; and then employed for demonstrating the necessarily ecologically-oriented practises of an authentic upper class household of the late eighteenth century – but still only as our mental depiction. If measured against the consolidated standards of conservation, the solution chosen is certainly unorthodox. There was a need for more space in order to be able to develop Gunnebo along the lines of the action plan. But, on the one hand, the building complex as now standing will give an impression of a household more economically resourceful and/or even more aesthetically committed than really was the case. On the other hand, a reconstruction of what had actually been the servants' building – that is, the farmhouse refurbished by Hall – was not possible because of lack of reliable documentation. In this situation a contemporary building would, according to established conservation principles, have been the obvious solution. But such a solution would have been less attractive when taking the ambiance of the place into account, and the ambitions of the action plan to re-establish and develop the eighteenth century character. Furthermore, the rewarding exercise in exploring and explaining elements of eighteenth century building technology as well has training craftsmen in that revived technology had not been possible with a conventional solution of that kind. The solution decided upon presupposes and demands that visitors are supplied with explicit information about the character of the 'reconstruction' and the objectives behind the decision to realise this unconventional approach.

Other (Re)Constructions: the Kitchen Garden, the Green House, the Farm Buildings and the Formal Park

The kitchen garden had been in regular use up until the 1940s, when Gunnebo was purchased by the municipality of Mölndal. In fact, some individuals still grew vegetables and potatoes right up to the start of the reconstruction work. As indicated above, a functioning kitchen garden was one of the key factors in the revitalisation concept laid out in the new plan for Gunnebo. Hence, a reconstruction was necessary in order to enable the visitors to gain insight into the interdependence between such a kitchen garden, the formal garden and the farm. In preparing the reconstruction, an analysis of the site-layout of the former kitchen garden disclosed that this area had been connected to the formal garden by sight-lines and axes like garden paths. This discovery only served to underline the importance of reconstructing the kitchen garden. The preserved designs of Carlberg allowed a reconstruction in great detail: vegetable patches, espaliers and hotbeds were laid out just as Carlberg had planned them.

In order to re-create the gardens of Gunnebo in an authentic fashion, it was necessary also to reconstruct the greenhouse for the purposes of growing plants and making it possible for exotic plants to survive during wintertime in the Swedish climate. In the case of the greenhouse, only a few sketches from early nineteenth century had survived. However, these sketches, extensive archaeological excavations on site, and comparative studies gave a general idea of the original greenhouse. In this case, a true reconstruction of a previously existing building would have been possible had sufficient documentation been available. However because of the lack of such documentation, the greenhouse also turned out as an 'imaginary reconstruction', but one which does neither aim nor claim to be a true copy. Nonetheless, it serves its function well and is appreciated by the public as a beautiful and interesting building.

The existing relatively modern cowhouse (from mid twentieth century) was partly renovated in order to house some live stock consisting of older breeds including a horse used for transport on the estate and sheep providing lamb for the restaurant and grazing to keep an open landscape. New buildings for sheep and hens were also built. It is planned to restore an old forester cottage in order to serve as a base for activities directed at children.

The formal park had been partly restored already in the 1950s. A new restoration project was initiated in 1994, designed by the landscape architect Kolbjörn Waern. In principle his proposal fulfilled the ambitions of the 1950s. The top priority of this project was to reconstruct the fences and parapets defining the surrounding space of the main building. Even in this case, Carlberg's own designs could be followed. This has also been the case with regard to other details that once were found in the park, such as trellises and urns. Extensive archaeological excavations were now undertaken in order to re-create the original spatial structuring of the site – an important feature of a park from the eighteenth century. These excavations not only allowed for the reconstruction of such features in detail they also proved that Carlberg's design actually had been followed. The ambition to recreate the original plant material of the park was somewhat more difficult to achieve, since only the exotic plants kept in the orangery had been registered. Contemporary literature and seed catalogues have added information on what might reasonably have been cultivated. It should also be mentioned that one element of modernisation was allowed for: the introduction of a sprinkler system to facilitate the maintenance of the park.

Problems Ahead

Today, the most important items of the action plan have been realised, and the preconditions for a qualified presentation of Gunnebo as a heritage site, as well as for the reception of more visitors are completed. The main building has been adapted so that a slightly higher number of visitors can be accommodated. The formal garden has been restored and the first part of the reconstruction of the kitchen garden has been finished. The 'reconstructed' servants' building houses a coffee shop and restaurant with a reasonable capacity and fairly wide selection of food and drinks, drawing on products from Gunnebo's own farm and kitchen garden. The reconstruction project itself has attracted media attention, and this has led to an ever growing number of visitors. Counting visitors to the main building only, these have grown in numbers from 3,000 to 20,000 a year. This very positive result is now creating growing problems in counteracting the preservational and educational goals that were the very motivation for the action plan. For example, the main building must be protected against being overused in order not to be destroyed. Public access for pedagogical purposes should thus, according to the goals, be prioritised and the municipality is invited to accept restrictions in their use of the building for formal receptions, where exclusive ambiance is the sought for quality of this the municipality's most distinguished and ever more popular building. Instead, as a result of the project, Gunnebo has become one of the most distinguished places for high level meetings in the region. Gunnebo was typically the meeting place chosen when the EU recently met with the American President.

Another problem, which unfortunately is much more difficult to solve, is the tension between the municipality's decision that the restaurant itself should carry the maintenance costs of the servants' building, and the fact that the restaurant is confined to the limited space offered by the 'new original' building. This, of course, has the consequence that the number of people that can be served is restricted. In this situation, the restaurant is mounting pressure to be permitted to rebuild and enlarge its occupied area in order to expand the number of seats, and thus the economic margins. That would mean installing offices in the cold attic, left in the reconstruction in this state as part of the educational programme. Such a change would thus again compromise the very goal of the reconstruction. Today the operation of Gunnebo is converted into a municipality owned company, which makes the economic management even more important. There are now also discussions of reconstructing more buildings, since the ever more popular restaurant needs more space to increase the economic margins.

This case exemplifies a major challenge for the future of Gunnebo namely how can a balance be achieved between modern economic principles of self-sustaining activities housed in spaces under conservation protection, and the educational and cultural ambitions that were the very reason for the reconstruction and conservation scheme? This is certainly a classical challenge in an age that simultaneously attempts to commodify the heritage and conserve its pristine authenticity for both future and present pedagogical purposes.

References

Baeckström, A. (1953), 'Gunnebo Rediviva', *Byggmästaren* A5:1953.
Baeckström, A. (1977), *Gunnebo I-III*. Stockholm: Nordiska museets handlingar 89 Göteborg. (Part III contains Carlberg's drawings.)
Bauer, W. (1956), *Förslag till rekonstruktion av trädgården* Havekunst 2:1956.
Ernstsson, I. (2000), *Gunnebo åter till 1700-talet.* Proceedings from a conference on garden history. Museum of Bergen. Annual report (Summary in English).
ICOMOS (1964), 'The Venice Charter'.
Jonsson, P. and Tollbom, A. (1981), *Gunnebo slott 1981: skadeinventering och förslag till plan för reparationer.* Not in print.
Strömberg, H. (1977), *Carlberg och Gunnebo: drömmen om ett slott*. Mölndal. Not in print.
Waern, Kolbjörn (1994), *Gunnebo - Förslag till åtgärder i parken*. Göteborg.

9 Adaptive Re-use of Historic Properties: Wollaton Hall and Park, Nottingham

ANGELA PHELPS

Figure 9.1 South façade of Wollaton Hall showing part of formal garden with original pond
Picture: A. Phelps 2001

No map can do justice to the siting of Wollaton Hall. Crowning a rise in the Sherwood Sandstone it sits above the surrounding country, in clear view from the centre of the city of Nottingham (Figure 9.1). One can only wonder at the boldness of the men who added the glass confection of the High Hall to further emphasise the process of viewing. Wollaton Hall and Park is an example of a particularly English category of built heritage: the ancestral seat of an aristocratic family, which once provided both wealth from an estate and also graceful living in impressive surroundings. Although some reject the oxymoron 'stately home', preferring the term 'country house', most would support Mandler's proposition that the survival of these properties promotes a conceptual illusion of Englishness: 'they epitomise the English love of domesticity, of the countryside, of hierarchy, continuity and tradition' (Mandler, 1997:1). The most notable feature defining a country house is substantial size, which is also the main reason that most have lost their value as homes in the modern period. Although by appearance and usage these properties carry an aspect of the heritage that may justify retention, their survival is not assured.

The purpose of this chapter is to explore the options for properties such as Wollaton Hall and to consider their potential for development as visitor attractions. The nature of the resource will be described, with particular reference to the house and its architectural merits. The current situation regarding public use and access will be explored. Finally, the recommendations of a recent conservation review will be considered in the light of changing attitudes towards the presentation of built heritage.

Wollaton Hall and Park: the Resource

Wollaton Hall is notable as a survival of one of the great achievements of Tudor architecture, now listed Grade 1 in recognition of its national importance for its early date and flamboyant style. The Tudor and Stuart period as a whole saw substantial investment in building, reflecting a marked improvement in standards of living throughout society. Malcolm Airs comments that 'the surviving evidence for such unprecedented architectural creation is visible throughout much of England and provides a unique contribution to our cultural heritage' (Airs, 1995:3). However, it provides also an example of the complex problems faced in orchestrating adaptive re-use to ensure survival for future generations.

Wollaton Hall is one of a group of houses representing the earliest manifestation of the conspicuous display that became associated with the concept of an English county house. The quality of styles pre-dating the more familiar classical architecture of the seventeenth and eighteenth century have only recently been re-appraised. 'It is now generally accepted that the English country house in the period between 1500 and 1640 was the product of enthusiastic and genuine architectural creativity and a worthy contribution to the culture of a dynamic age' (Airs, 1995:xi). Built between 1581 and 1588 Wollaton Hall reflects a period when the relative peace of the Elizabethan age encouraged the development of an architectural competition that saw the construction of ever grander houses. Houses such as Wollaton Hall, Hardwick Hall, Knole and Petworth were built for the aggrandisement of their owners: they were built to impress, and often to entertain, Queen Elizabeth I, her courtiers and others of noble birth. These houses have been the object of envy from their earliest conception. As such, the impression given to visitors was often more important to their owners than the comfort they provided for residence. Consequently the houses that have survived have either been abandoned for long periods with their owners living elsewhere, or have been substantially altered by subsequent generations to create more comfortable accommodation.

Wollaton Hall was commissioned by Sir Francis Willoughby (1546-1596). Sir Francis was born into a wealthy and well-connected family, although compromised by involvement with the royal aspirations of the Greys of Bradgate Park, Leicestershire. Sir Francis was orphaned at two; as a younger son, he was over-shadowed by his brother, and was sent away for education at schools from an early age. Sir Francis inherited the estate when he was just 18, on the unexpected death of his brother; some historians claim he retained a sense of inferiority within a social system of strict hierarchy, possibly accounting for the luxurious lifestyle he adopted. He was presented at court in 1575, an occasion which may have been expected to be followed by a royal visit to his estate. Queen Elizabeth was well known for her travels and her courtiers competed to entertain her. However, the Willoughby family lived in an old manor house in Wollaton village, unsuitable for the vast retinue accompanying a royal progress. If Sir Francis Willoughby aspired to display his nobility he would have needed a substantial house to attract royal interest. 'New features, such as a deliberately contrived symmetry, or vast expanse of window, drew attention to the builder's modernity and the prosperity of his estate' (Airs, 1995:4). Wollaton Hall displays these features in the most exaggerated style, with medieval references that suggest a deliberate attempt to underline the ancient ancestry of the family.

Wollaton Hall is attributed to Robert Smythson, noted as the first person in England to be publicly ascribed the title 'architect'. One factor contributing to the importance of Wollaton Hall is that considerable documentary evidence survives from the early phase of both building and occupation, including building accounts and room inventories from Tudor and Stuart periods. Pamela Marshall has written an extensive account of the building and household in two monographs published by Nottingham Civic Society. She stresses the historic importance of Wollaton Hall as a significant social and architectural transition: 'in late sixteenth century Nottinghamshire Wollaton Hall must have stunned its beholders, few of whom could have seen anything remotely comparable, yet internally the house was designed to work according to the strict conventions of late medieval domestic planning' (Marshall, 1996:5). It is estimated that the house cost some £8,000 to build, at the outset manageable within the annual revenue of the coal mines on the estate. However, coal prices were falling, and Sir Francis' family expenses rose with the marriage of three daughters and a separation settlement for his estranged wife. In the latter part of his life Sir Francis chose to run up debts rather than sell land or curtail his lavish spending; thus he compromised the family fortune, leaving financial problems for generations to follow.

The fabric of the house was damaged by fire in 1642, and after the death of Sir Francis Willoughby's heirs in 1643, the huge and partially derelict house remained empty for a generation. The house was reoccupied in 1687 by the forth Sir Francis Willoughby and his sister Cassandra.[1] Much has been made in recent interpretation of a rather tenuous connection with their father, the third Sir Francis Willoughby (1635-1665), the famous naturalist. He inherited the property shortly before his early death but never lived there; nevertheless a natural history association was created when his scientific collection was moved to the Hall and catalogued there by Cassandra. Repairs were made and part of the house was remodelled to facilitate family living. In the early nineteenth century Sir Thomas Willoughby, by then 6th Baron Middleton, commissioned Jeffry Wyatt (subsequently Wyatville) to undertake more substantial alterations, which finally removed the Tudor interior. Wyatville was developing a reputation for 'restoration' projects, reflecting both the rising aspirations and increasing wealth of the landed gentry. Such alterations would not fall into the modern understanding of restoration, being concerned with creating fashionable new interiors rather than going back to previous styles. Wyatville used mezzanine floors to create more rooms within the house to provide separate accommodation for servants. The cramped, poorly lit conditions of the rooms created in this period are in strong contrast to the open, airy, well lit and large rooms of the original design. Most of the Tudor interior was removed and replaced with Georgian plaster cornices, ceilings and door frames, achieving a reduction in the apparent size of the windows. The most obvious structural change was in the entrance, where Wyatville created a lobby leading directly into the Great Hall. Fortunately for current interests, he left the Tudor screen that encompassed the original entrance to the Great Hall. Wyatville also designed a single story extension on the west side to house the growing servant needs. For the following generation, a substantial courtyard building was constructed to accommodate stables and other servant functions.

Despite expensive work on the Wollaton estate the family moved to Yorkshire in this period and the house was occupied less, eventually being let to a tenant. In 1920 the 11th Baron Middleton sold Wollaton Hall, its garden and parkland to Nottingham Corporation (now Nottingham City Council) to raise funds to cover death duties. At the time of the purchase, the main consideration of the Corporation was to acquire land for building houses and relief roads. A study of council minutes and local papers at the time of the sale demonstrates that although local people raised questions concerning conservation and public access to the Hall and Park, these issues were not discussed by the council (Pearson, 2000). Consequently it is of little surprise that the property passed into public ownership with no assessment of its state of repair, continued use or conservation needs. Although the Park has been managed as a public asset, the Hall was seen as a liability.

There was no obvious use for the Hall, which had been acquired without its contents. The Corporation had in its possession, but in store, a substantial natural history collection. It is not so surprising that the empty Hall was seen as an opportunity to solve two problems at once, by moving the collection into the Hall. At the time there were many historic houses still in occupation, so adaptation to alternative use aroused less public interest than it may do today. The conservation of the specimens in a natural history collection requires controlled light, so the interior of the house was converted to museum conditions. Although this was dictated by curatorial needs, prioritising the collection obstructs the view of the house and disastrously undermines appreciation of the most distinctive features of its architecture – light and air. Thus Wollaton Hall became a visitor attraction, opening as a Natural History Museum in 1926. It still functions as such today, attracting some 100,000 visits in 1998 (English Tourism Council, 1999). This number is well below Nottingham Castle, the most popular local museum, which attracted some 330,000 visits in the same year. However the figure is considerably more than the other historic house owned by Nottingham City Council at Newstead Abbey (c. 30,000).

The grounds comprise a substantial enclosed park with lake, gate lodges, a walled nursery garden, formal gardens and the courtyard buildings (Figure 9.2). The park is stocked with red deer, roe deer and white cattle. The red deer are claimed to be direct descendants of the deer of Sherwood Forest, trapped by the original enclosure of the park. Wollaton had been famous for its herd of native white cattle, but these were disbanded in 1825; in 1988 a small herd of white cattle was re-introduced to recreate the parkland scene. The original garden design is of great historic interest; a surviving sketch from Smythson's plan shows that the symmetry of the house was to extend into the surrounding grounds – not an easy plan to execute given the topography. Gervase Jackson-Stops considers the garden at Wollaton Hall to be the first English example of a garden deliberately designed to relate to the house, in this case with eight garden squares mimicking the symmetry of the house: 'few other Elizabethan houses were planned as an architectural unit' (Jackson-Stops, 1988:14). It is uncertain how much of the plan was ever completed; the only trace in the current garden is the circular pond (Figure 9.1). Although such a garden is of significant interest today, it was given little consideration at the time the Corporation purchased the property. The park was much reduced in size by land sale for both road and house-building schemes; part of the remaining park was leased as a golf course, and further enclosures made to retain the deer and rare breed cattle. Today the park provides informal recreation space and is used for festival events. The 'parks' management is evident in the incorporation of car parks, sport pitches, seats and Victorian-style mass bedding in the gardens. Despite considerable destruction the surviving garden and park remains sufficiently important to be listed Grade II. There is no accurate measure of the number of visitors who use the park, or recent survey to reflect their origin and interests.[2]

Figure 9.2 West façade of Wollaton Hall showing nineteenth-century extension for servants' quarters and Courtyard buildings
Picture: A. Phelps 2001

The courtyard buildings have been put to multiple uses. Nottingham City Museum's industrial collection is housed here, but the museum has experienced periods of closure due to cost-cutting measures in the late 1990s. The most recent adaptation is a gallery for temporary exhibitions, with a small shop and wildlife interpretation room. Other space is used for storage and leased as offices. Catering is franchised, using a poorly situated temporary building; in 2001 the café was closed on health and safety grounds. The courtyard buildings are under-used and falling into disrepair. Despite the presence on-site of stonemasons, there is insufficient investment to achieve more than essential repair. The site presents a dilemma; although the park is well used, only a small proportion of visitors climb the hill to the Hall. There is greater local interest in dog-walking and occasional events than either the architectural heritage of the Hall or its current museum use. For some years funding has been insufficient to make any radical alterations to management, and there is little prospect of material change within the Local Authority budget. There is no shortage of ideas about redevelopment, but to date, no clear strategy for funding new investment.

Historic Properties as Heritage Attractions

The idea of visiting a country house is not new, but the concept of supporting the upkeep of a house through creating a 'visitor attraction' is. Raising sufficient revenue to fund maintenance requires a substantial business, and funding major conservation work is probably beyond the reach of even the most successful tourism venture. The private houses that have become successful visitor attractions in England are in the minority, and share both a long history of public access and a management committed to providing both a quality experience and new attractions to promote repeat visiting.

In England there are three main categories of country house available as visitor attractions:

- properties in the guardianship of organisations such as English Heritage and the National Trust that have conservation as their primary purpose (e.g. Kedleston Hall, Calke Abbey, Brodsworth Hall);
- properties in public ownership, with access obligations but no conservation strategy (e.g. Wollaton Hall, Elvaston Castle);
- properties in private ownership (e.g. Chatsworth, Longleat, Beaulieu).

In the first case the conservation needs of the property will take priority; in the latter case the commercial interests of the family may take precedence. Properties in public ownership are in a more perilous situation, with funding subject to political competition, few resources for speculative development and councillors who may see access as a social priority.

Opening to the public is not a simple matter. In the first place the property has to be suitable to attract and retain an audience, within an increasingly competitive market. This requires a good location, either already frequented by tourists (predominantly in and around London for overseas visitors, the south coast and south west for domestic tourism) or within easy reach of a substantial day-trip market (central England or the home counties and Midlands). The more obscure locations require powerful incentives to encourage sufficient people to travel for a successful business (e.g. the theme park at Alton Towers). With the possible exception of major properties within reach of volume international trade (e.g. Blenheim), attractions need to be regularly improved and augmented to generate repeat visits (e.g. new rides in the grounds, opening rooms not previously accessible). Many properties resort to special events to attract a wider social range of visitor. Even the organisations protected from the main pressures of commercial tourism promote special events to enhance visitor numbers: English Heritage publish a varied calendar of historic re-enactments, theatre and music at their properties, while the National Trust favours summer concerts, craft fairs and plant sales.

The management of properties like Wollaton Hall through public bodies presents further difficulties. A local authority will hold a large and varied portfolio of buildings, within which management is likely to prioritise current function rather than architectural or heritage merit. In the case of Wollaton Hall the management of the park became the responsibility of the estates department along with other urban parks across the City, while the Hall and courtyard buildings were placed in the care of the museum service, an arrangement that has only recently been revised to introduce a site manager. As a local authority the Nottingham City Council is answerable to councillors and ratepayers for finding 'best value' in its resources. Inevitably there is likely to be greater concern for short term expediency than with preservation bodies such as the National Trust. Any property in such ownership is subject to adaptive re-use to justify continued maintenance. The challenge with properties such as Wollaton Hall is to find uses that satisfy the needs of earning income, provide public access but also do not drastically undermine conservation interests. The solution found in the 1920s has provided its own legacy into the current debate, as any alternative strategy for the Hall will require the relocation of the important natural history collection.

There have been various attempts in the last two decades to find alternative solutions that may improve both the conservation and presentation of the Hall. As time passes, the need for substantial maintenance increases. In 1993 a Management Plan for Wollaton Park was drawn up which prioritised 'environmentally sound' management: 'in this enlightened age management should be geared towards maintaining the living landscapes rather than seeking to subdue it' (Mayfield, 1993:18). Nevertheless this plan envisaged developing visitor facilities, particularly an attraction in the walled garden, an environmental resource centre and new refreshments located in the courtyard buildings. However, this report recognised that little could be done in the Hall so long as its use as a museum continued. At this time there was another redundant building in Nottingham prompting debate about re-use. The Victorian brick-built Low-Level station was proposed as a new location for the natural history collection. This plan never materialised: after much enquiry the station site was sold to a property developer and listed-buildings consent was granted for conversion to a commercial leisure attraction housing a multi-screen cinema, restaurant and bars.

The 1993 Management Plan also presented the debate concerning a proposal to designate Wollaton Park as a Local Nature Reserve. This was rejected on the grounds that such status would not provide any further protection than that existing within the Planning Authority's Conservation Area Status, but might place restrictions on management regarding public access. As a Local Nature Reserve management would be required to prioritise the needs of wildlife conservation; despite the 'green' tone of the Park Management Plan, it was conceded that the City Council wished to retain the right to prioritise provision for visitors. Nevertheless the Plan also includes an objective to publicise other facilities in Nottingham managed by the Council, in an attempt to prevent the 'over-use' of Wollaton Park.

Towards the end of the decade it seemed a new opportunity had been presented by the National Lottery. This prompted a further review of the site, with funds provided for a professional conservation strategy. The consultants' review identified the following objectives (Porter & Williams, 2000):

- enhance visitor experience by improving the quality of offer and increasing public access;
- improve the conservation of the Hall and Park;

- improve the interpretation of the Hall and Park;
- improve the amenity value of the Hall and Park;
- encourage environmental improvements and sustainable developments;
- open services to public view and involvement through regular events;
- improve accommodation of, and access to, collections;
- remove and relocate all materials and uses/services that do not relate directly to or support the Wollaton estate.

The report presents a 'vision' that 'builds on the concept of Wollaton as a vibrant working estate': 'Wollaton will be unique in the way that the history, present and future landscape, buildings and wildlife will be treated as a single entity' (Porter & Williams, 2000: 2). This report reflects the recent shift in Local Authority thinking towards social inclusion and access, and also reflects the shift in practice of English Heritage towards retaining complexes rather than individual buildings.

A substantial capital grant would be necessary for the refurbishment. The review envisages raising revenue by improving the quality of catering, to include the possibility of private events and corporate hospitality. Others question whether this is achievable within the suggested location of the lower ground floor servants' quarters. The Courtyard buildings provide a much better opportunity to tap the wider visitor population, as it straddles the route from the car parks to the popular walks around the garden and lakeside. However, suggestions to improve the presentation of the Hall are frustrated by the existing museum conditions. The City of Nottingham controls eight museums with varied collections; reduced budgets render it increasing difficult to maintain access to the collections and accommodate the new council initiatives in social inclusion. Without a radical review of the entire museum function, a major constraint on redevelopment at Wollaton will remain.

Whatever the preferred plan, application to the National Lottery is a time-consuming business. While debate within the Council continued the national strategy for the lottery has changed. The very substantial grants to individual projects of the early years have been curtailed, with greater focus on spreading the benefits through more, smaller awards. The Council still favour proposals to develop the site as a more substantial tourist attraction, but it is questionable whether sufficient funds will be forthcoming to allow significant conservation of the buildings and the major re-organisation of the museum collections suggested. Without such investment the fabric of the buildings will continue to decline, as current funding is inadequate to undertake major conservation programmes.

Conclusion

Despite decades of destruction, England still has a wealth of country houses. Most of these are now protected by the listing system designed to identify and protect architectural excellence. However, this system has created problems for the future of the buildings, as it becomes increasingly difficult to find uses that will produce the revenue needed for maintenance. Wollaton Hall demonstrates a particular problem brought about by the circumstances of the period in which it changes hands. Had the Willoughby family been forced to give it up sooner, it may well have been demolished. Had they held on for another couple of decades, it is likely the Hall would now be held by the National Trust. However, the sale in the 1920s transferred the property into the hands of a local authority that had neither the interest nor the resources to maintain it as a house. The Hall has survived, but as a shell for a museum collection. Although attitudes within the Authority are now very different to the time it was acquired, the current museum use of the Hall is creating a substantial block to any alternative future.

An important question concerns the conservation priorities. If the purpose of conservation is to retain the building, then with some additional funding to improve maintenance the existing use could be allowed to continue. Although the Hall is not ideal as a natural history museum, it does at least provide space. A major factor that mitigates against any substantial change is the loss of the interior contemporary with the exterior. With neither Tudor fittings nor furnishings it would be a vain hope to return the Hall to its earliest glory. Even if this were possible, there is no evidence that sufficient visitors could be attracted by the presentation of house and family to support it as a commercial visitor attraction. Although National Trust properties within the region attract significant numbers, they are only open for half the year and rely heavily on volunteer staff. Nottingham does not attract large numbers of tourists, so the audience needs to be drawn from the local population, requiring an attraction that will support repeat visiting. One alternative would be to continue museum use, but to change the collection to one that does not need the careful control of environment required for natural history. Ceramics and decorative arts would tolerate more light, allowing the opening of windows to restore some of the style of the interior. However, any such use would not generate a significant increase in the visiting public. More visitors could be attracted to the house if its rooms were cleared and used for activities – educational workshops, concerts and temporary exhibitions as suggested in both recent reviews. However, the building is not suitable for large events as the access is poor and there are many steps. The greatest opportunity for income generation lies in the current visitors to the park, if appropriate services could be provided more effectively. However, any radical change in use would require a substantial input of resources, unlikely unless a lottery bid is successful.

The recent management plans for Wollaton Hall and Park reflect the complicated context of local authority ownership. Local Authorities are responsive to a range of pressures ranging from national political policies to the desires and needs of the local population. In the last decade interests have shifted from an awakening of environmental consciousness to the demands of social inclusion. Plans prioritising sustainable development and biodiversity have given way to arguments for enhanced intellectual and physical access. These need not be mutually exclusive, but in the context of declining public funding and sharper accountability, the increased costs of meeting both agendas creates formidable obstacles. So long as properties such as Wollaton Hall are held by local authorities their conservation will hang in the balance of multiple responsibilities. Inadequate funding to achieve the quality of outcomes desired too often results in a management vacuum in which the conservation needs of the properties are neglected.

Notes

1. Sir Francis Willoughby, the builder of Wollaton Hall, left no male heir. On the death in infancy of his only son, he arranged a marriage between his eldest daughter Bridget and his second cousin Percival Willoughby. Percival was the grandson of Sir Francis' father's sister; the co-incidence of surname no doubt presented the match. The family history is recounted in detail by Pamela Marshall in her monographs for Nottingham Civic Society.
2. From February to July 2001 Wollaton Hall and Park were closed due to the national outbreak of foot and mouth disease. Despite the Park being entirely surrounded by the urban area of Nottingham, the Council determined that the protection of the deer and cattle was sufficiently important to prevent public access.

References

Airs, M. (1995), *The Tudor and Jacobean Country House: a building history*, Alan Sutton Publishing, Stroud.
English Tourism Council (1999), *Visits to Tourist Attractions*, HMSO, London.
Jackson-Stops, G. (1988), *The Country House garden: a grand tour*, Pavilion, London.
Mandler, P. (1997), *The Fall and Rise of the Stately Home*, Yale U.P., Newhaven & London.

Marshall, P. (1996), *Wollaton Hall: an archaeological survey*, Nottingham Civic Society, Nottingham.
Marshall, P. (1999), *Wollaton Hall and the Willoughby Family*, Nottingham Civic Society, Nottingham.
Mayfield, M. (1993), *Wollaton Park Draft Management Plan*, Nottingham City Council, Nottingham.
Pearson, K. (2000), *Wollaton Hall* unpublished MA thesis, Nottingham Trent University.
Porter, G. & Williams, J. (2000), *Wollaton Hall & Park: Outline strategic plan*, Nottingham City Council, Nottingham.

10 Managing the Heritage of Fortress Towns: the Cases of Naarden and Bourtange

G.J. ASHWORTH and M.J. KUIPERS

The single most prominent and inescapable reality of the past 1000 years of European history has been the existence of almost continuous chronic internecine conflict between dynasties, nations, ethnicities and classes. This has left a legacy in military architecture throughout the continent. Almost all continental European towns were at one time walled and these defences remained until well into the nineteenth century and in some cases much later. Although most such defence works have subsequently been demolished for the value of the land they occupied or in order to remove barriers to traffic circulation, many remain. These now provide dramatic heritage structures and often high profile tourism attractions. However they also provide a distinctive if widespread heritage management case. The common features possessed by such towns is recognised by the establishment of an international forum, 'The European Walled Towns Friendship Association' which consists largely of smaller purpose built fortress towns with intact fortification systems and an interest in utilising the heritage potential that this endows (Bruce, 1999).

A distinction is made here between walled towns, that is multifunctional towns walled for their own protection, and fortress towns, where the defence function existed for supra-local reasons and where the military function was the *raison d'être* for the existence of the settlement. It is the second category that provides the most distinctive and thus interesting cases for consideration here.

The Characteristics of Fortress Towns

Fortress towns have a number of distinctive characteristics relevant here and are archetypal examples of gem cities. These are fortuitously preserved settlements with a complete integrity in both their spatial comprehensiveness and single historical period. Their very perfection raises both possibilities and problems.

Physical Structure

In military terms, defence works were merely a means of purchasing a local tactical advantage over an attacking force making defence easier. Because defensive walls were always expensive to construct and to maintain, their length was kept to a minimum, which favours the enclosure of a circular town as compactly as possible. Similarly, military necessity imposes restrictions upon the expansion of the town beyond the walls, even in times of peace, and usually strict controls on not only building but also planting in the fields of fire and approaches. For these simple imperatives fortress towns were physically compact and have tended to remain so longer than other settlements. Related to this and also to the paramountcy of the military function, urban development and diversification was not only hindered by the accessibility constraints but usually was also discouraged by military authorities.

A further constraint hindering non-defence development, even when obsolescence and demolition of the fortifications has occurred is the nature of the spatial location of fortresses. The location and site characteristics were originally chosen for their defence features, rather than any other advantage, which makes subsequent functional diversification more difficult. In addition fortress towns were often sited in peripheral borderlands isolated from the centres of economic and political power whose frontiers they were intended to defend.

The compactness, circular character and obvious physical architectural encirclement does however convey an image of the city that is easy to grasp even on short acquaintance. The converse of this is that the completeness and geometrical complexity of the fortress and its architecture is generally difficult to visualise on the ground. Indeed it is often only properly visible from the air as most postcards and publicity material of fortress towns readily testifies.

Finally it can be argued that the physical characteristics of containment and isolation may be reflected in social attitudes of the population that continue even long after the disappearance of the military function. The population of garrison towns tended to be socially cohesive but was essentially a foreign implant with few links with the wider region outside the town. The physical containment and isolation, as well as the nature of military life, may be reflected in social characteristics of an inward looking conservative defensiveness against the enemy without. Such an environmental determinism may seem unlikely but the fact that such arguments are used regularly in current policy documents suggests that whether correct or not such views have an impact upon development potential and policy.

The Tourism

Walls enclose and bound the tourism product, allowing easy orientation for the visitor and frequently a wall top walkway allowing a circular tourist routing. The tourism product and its location are clear, unambiguous and can be almost instantly appreciated. The main disadvantage for tourism development stems from the same clarity and completeness. It is very difficult to vary or extend the product by offering new tourism experiences or even tourism support services, which might distract or conflict with the defence heritage on offer. Thus visits tend to be very short and the fortress is generally consumed within an hour or two. It is thus difficult for fortresses to either be profitable tourism complexes in themselves or develop in such a way to achieve such profitability. In addition they are, by the nature of the defence function, frequently located on peripheral frontiers distant from the main centres of demand and inherently difficult of access.

The Heritage

In terms of the heritage itself it is worth remembering that walls were built to keep enemies out or contain and control untrustworthy populations. It is a memorial to military, police, customs or taxation functions; none of which would ostensibly seem to favour the development of entertainment products. Fortresses historically represent national or dynastic enmities and conquests or local tyranny and oppression. They therefore would seem to be especially dissonant in many respects. Curiously the wall can also be interpreted as a symbol of friendship, networking, containing and accommodating visitors. More prosaically fortress towns because of their completeness and thus necessity to sustain the integrity of their physical structure are expensive to maintain and difficult to use for modern purposes. The obvious question this raises is thus how many are needed?

The Case of Naarden

History

There are indications that Naarden has been fortified from the tenth century. Originally sited on the old Zuider Zee it received municipal rights between 1321 and 1337, and became the capital city of *Nardincland*. At that time, it probably was surrounded by a simple palisade but was destroyed by the *Hoekse benden*, in 1350. The city was rebuilt, relocated and refortified eventually with stone walls and a moat. Its moment of historical significance occurred in 1572 when, early in the Dutch revolt, it was invested, taken and its population massacred by Hapsburg forces. The resistance and fate of Naarden became a powerful symbol in the founding mythology of the Dutch Republic. The fortress was reoccupied and rebuilt by the States of Holland between 1573 and 1579. The fortress designed by Thomas Thomasz and Adriaen Anthonisz was the basic form of a low wall, 'wet' *gracht* and six artillery bastions (Stichting Menno van Coehoorn, 1964).

Subsequently the fortress was occupied by the French, recaptured in 1672 and rebuilt between 1672-1685 to its present form (see Figure 10.1) with bomb-proof gates, bomb-free barracks, casemates and shelters by Adriaan Dortsman and Willem Paen.

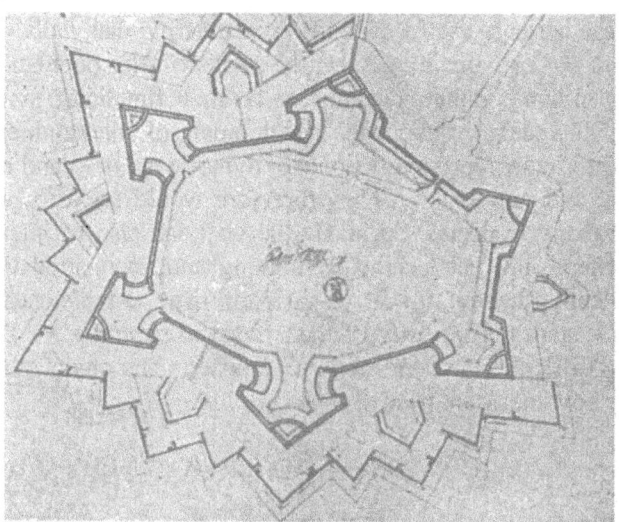

Figure 10.1 Design for the rebuilding of the fortress of Naarden in 1679 (With the signature of the Prince of Orange III)
Source: ALGEMEEN RYKSARCHIEF's-Gravenhage

Although it was never again attacked its military function continued, largely as a result of its strategic location. It was not abandoned as an artillery position until 1915 and decommissioned as a fortress in 1926, although it continued to be part of the 'New Dutch Defence Line' of fortified positions and inundatable areas until well into the twentieth century. However its heritage value was early recognised, perhaps as a result of its role in the 80 years war, and it was placed on the 'Temporary List of Dutch Monuments of History and Art' in 1921.

Modern Developments

The long military-strategic function of Naarden, its main reason for existence, lasted long enough for its historical value to be appreciated by the time it was redundant. There was never the temptation to demolish the walls and fill in the ditches. This was probably because the town did not have or ever had other significant commercial or industrial functions nor was it well sited to develop substantial local market or trading functions.

The abolition of the active military function in 1926 did lead to some dilapidation but by 1932 the *Stichting Menno van Coehoorn*, a voluntary pressure group, had been created nationally by those recognising the historical value of old fortifications in The Netherlands. However it was not until 1953 that a local pressure group, the *Stichting Vrienden van de Vesting Naarden*, was founded: composed of interested people from outside as well as inside the town. At that time the structures were owned by the Ministry of War, which only maintained the buildings on those bastions that were in use, leaving the rest to fall into disrepair (Scheltema-Vriesendorp, 1990/1991). In 1961, the fortifications were transferred to the responsibility of the Ministry of Culture, Art and Science and the Department of Public Works was charged with renovation. The whole renovation process lasted 25 years and was not completed until 1990 (Scheltema-Vriesendorp, 1990/1991). In the period 1964 to 1977 the *Grote Kerk* in the fortress was renovated, the sewage disposal system renewed and the whole inner city of the fortress was re-paved. As a consequence of these improvements the fortress became more attractive as a residential environment and many domestic premises were renovated.

Place of Heritage and Tourism Now

Today the fortress of Naarden has largely a residential, shopping and tourist function. The maintenance of the walls and the other fortifications has stimulated two main developments.

First, it formed a major attraction for visitors, including tourists interested in heritage. This in turn has led to the growth in some tourism orientated services including three museums. In the last five years particularly, a sort of cultural-historic complex has been created identifiable in the character of the shops and events and festivities that are organised in the fortress. An example of this is the location of the studio *Jan des Bouvrie* in the arsenal of the fortress. It consists of shops concerned with interior design and also theatre, a boutique and a restaurant. The complex has become in itself a tourism attraction and has led to a considerable growth in the number of visitors in the past five years.

Secondly, Naarden has a unique residential atmosphere. Probably it is the physical intimacy of the fortress contained within the walls as well as the general atmosphere of historicity that has made it an attractive residential location for certain types of residents. These are people who place a view value upon residential environmental amenity, the monumental historicity of their houses and perhaps also, at least outside the times of tourist visitation, the tranquillity in the fortress. In the last few years the fortress has attracted many residents out of the higher social classes. As a consequence of this the housing prices in the fortress have risen considerably.

On the other hand the maintenance of the walls, the other fortifications, the many monumental buildings and the original street-pattern have imposed severe limitations on development, including upon tourism. Although there are enough parking facilities in the car park built on one of the bastions of the fortress, within a very short walking distance from the main street, the *Marktstraat*, it is scarcely used by the visitors. Almost all visitors park their cars in the streets of the fortress, and because the fortress is very compact and the streets are very narrow, this leads to congestion, especially on busy days, such as Saturdays and during holidays. This again causes problems such as annoyance among the inhabitants of the fortress and disturbance in the experience of the fortress by its visitors. The holding of special events is especially difficult in this respect.

According to the *Bezoekersonderzoek* (NHTV, 1993:5) almost 40% of the visitors were of the opinion that the fortress of Naarden had enough possibilities and tourism attractions to spend a whole day in it. However, about 80% of the visitors of the fortress did actually spend no more than four hours in the fortress (NHTV, 1993: table 2.1.11). There is only one small hotel inside the fortress and the difficulty for this, and other similar towns, is to generate economic returns from relatively short tourism visits. The local government exercises very strict controls on new developments within the fortress, restricting the number of cafés, restaurants and snackbars, because it is of the opinion that the fortress must not become a sort of funfair attracting *patat-toerisme* ('chip tourism'). This renders tourism development extending the length of stay especially difficult. Similarly the residential and other functions operate within the severe constraints imposed by the primacy of the preservation function. Decisions about potential new developments within the fortress will always ultimately subordinate local interests to the wider interests and values of a national and international monument and preservation and conservation management will take precedence over local development, even in matters of heritage tourism.

The Case of Bourtange

History

East of Groningen and west of the German cities of Munster and Oldenburg is an extensive area of badly drained peat, which has repelled settlement and channelled transport either around or through it on a few narrow causeways. The Bourtange Moor to the east of the river Ems and stretching 80 kilometres south of the Dollart estuary was, until the drainage enterprises of the eighteenth and nineteenth centuries, traversed by only one dry routeway, the narrow causeway through the morass at Bourtange. It does not require much strategic insight to conclude that three positions will be critical to controlling all east-west movements, namely around the north (Nieuweschans), south (Coevorden) and through the moor where the village of Bourtange now stands. Contrary to what might be expected however the position at Bourtange was not first fortified in 1580 to hinder a westward movement from Germany but conversely was constructed by the States-General of the United Provinces to cut-off Groningen and its surrounding area of the north-east Netherlands which had declared support for the Hapsburgs part in the conquest and annexation of the North by the Holland/Flanders based Orangist rebels.

However this strategy of interdiction was successful. The fortress withstood a four day siege by imperial governor Verdugo from Groningen and was thus instrumental in leading to the siege and conquest of the city of Groningen by States forces in 1594 and therefore the creation of the current border between Dutch and German sovereignties. Its subsequent 300-year history was as a bastion for defence of the Northern Netherlands against threats from the east. Such threats materialised on three occasions. In 1605 when Hapsburg forces threatened to move against it as part of a more general offensive. In 1665-1666, when troops from the Bishop of Münster, with a claim upon Groningen attempted to advance around the southern end of the moor and were defeated at Ter Apel in part by troops sallying southwards from the Bourtange garrison. In 1672 when the most serious invasion of The Netherlands (at least until 1940) was conducted by forces from France, England and, in the north, Münster. The fortress at Bourtange was invested, bribery of the garrison attempted, but ultimately the position was turned by the fall of Coevorden in the south which opened up the way to the city of Groningen, rendering Bourtange all but irrelevant. The city of Groningen then suffered its most notable siege and what has become its 'finest hour' is still celebrated by a public holiday in the city.

Although the position was never again attacked, it remained in front line commission for a further 168 years being decommissioned in 1850. The strategy of fixed defence lines was however continued for a further 90 years but using the series of 'waterlines' (of which the nearest was the Zuider Zee – Ijssel, more than 100 kilometres to the south-west) which effectively abandoned the defence of the northern provinces, which were not defended against German invasion in 1940.

A position that has been fortified from around 1580 to 1850, has understandably changed many times. It is worth remembering that Bourtange's 270 years of active defence service did not begin until at least 200 years after the first effective military uses of gunpowder. The first, and simplest, fortress of the 1580s was constructed (by Sonoy) on the Italian system but extensive rebuilding occurred at the beginning of both seventeenth and eighteenth centuries (following a Coehoorn design) in response to the increasing muzzle velocity of artillery leading to longer ranges, harder hitting power.

The weaknesses of Bourtange are not so visible or so impressive as its strengths. Bourtange was never a town supporting a defence function but a national fortress surrounding a dependent settlement. As with any defence expenditure in time of peace it is difficult to justify and necessarily unproductive, while in time of war its necessity is obvious but investment too late. A sinking water table, in the later years of the fortress, caused by improved drainage elsewhere, which was steadily drying out the morass, weakened the effectiveness of the water defences. From its first beginnings the fortress in theory was to be maintained and manned by the central government and the two northern provinces of Friesland and Groningen. The States government would not, and probably could not, maintain a large standing army and depended very heavily on foreign mercenaries. The problem was compounded in Bourtange by the isolated location, which rendered it an unattractive posting, and there was a minimal local population that could be used in emergency.

From Fortress to Village and back to Fortress

The decommissioning of the fortress in 1850 was a decision of the Ministry of Defence that could be implemented at the stroke of a pen: the physical dismantling of the enormous structures however was costly and complex. Nevertheless it was seen locally as highly desirable and the burden was assumed as a national expense. The land was reallocated from public to private ownership and re-divided in such a way that parcels included both walls and ditch so that earth from the ramparts could be used as fill for the ditches and the land returned to largely agricultural use (see Figure 10.2). The ending of its military role removed almost all urban functions from the settlement. Bourtange had never developed service activities other than those necessary for sustaining the garrison.

Figure 10.2 Aerial view Bourtange 1951
Source: Foto KLM AEROCARTO - Arnheim

The population of the local authority at the time of decommissioning was only 350 most of whom most lived in farms outside the walls. The fortress as a physical structure had thus completely ceased to exist with the exception only of some of the relict buildings which were put to civilian uses and the morphological pattern of the settlement itself. The key question therefore is why was it reconstructed, effectively in the period 1971-1982? What shift in values occurred over little more than a century that transformed what had been considered to be quite literally valueless into Bourtange's major economic asset and principle focus of identity? The two wider coincident contexts are the long term structural decline in agriculture and the growth in heritage as an instrument of local planning. The first presented a particularly acute problem in the northern Dutch-German borderlands which were marginal in both spatial and economic terms which became especially apparent in the course of the 1960s. The second seemed to offer the only alternative strategy to emigration and the decision to reconstruct was formally taken by the Local Authority (*gemeente Vlagtwedde*) into which Bourtange had been incorporated, in 1964.

It is worth noting the radical nature of that decision which in retrospect is difficult to appreciate let alone understand the motives behind it. Although much of the finance was national, making use of various conservation and regional development subsidies, the initiative was local. It may now appear self-evidently worthwhile but in the mid-1960s the heritage movement was in its infancy and there were very few precedents and probably none where such a small settlement attempted such a major work of reconstruction. The fortress, however interesting, is neither unique nor associated other than peripherally with notable historical events or personalities. The extant structures (notably the Captain's residence, barrack building, powder magazine and protestant church) are of local architectural interest rather than major historical or artistic significance. More of the defence works actually remained at Oudeschans and reconstruction strategies have not been pursued in the neighbouring settlements of Nieuweschans (see Chapter 12) or Coevorden, or indeed in the city of Groningen which was both more extensively fortified and played a more central historical role. Bourtange had indeed fewer potential resources and thus options, but equally the potential demand for any heritage product on this location was likely to be limited by its spatial peripherality and absence of any other substantial neighbouring attraction. There is therefore no question of clustering to obtain a tourism product important enough to sustain overnight stays or indeed visits of more than an hour or so. The economic gains are thus likely to be minimal and mostly accrue to services well outside Bourtange. Tourism may well be seen as no more than marginal windfall gain for a product intended to satisfy some national or local identity raising in turn the question, 'then why put it here?' Willemstad in North Brabant is an obviously similar and better located case while overseas, Fort George in Scotland or Louisbourg, Nova Scotia present very similar issues (Ashworth, 1991).

Who are the Visitors and What are they Sold?

In comparison with Naarden, Bourtange has few facilities other than the fortress, four small museums ('Synagogue'; 'Barracks'; 'Captain's Lodging'; 'Smithy'), a café and a souvenir shop. The product on offer to tourists and local day visitors is the fortress itself as military architecture and defence strategy. There is a stress in the museums on the 'everyday life' of the Bourtange garrison and community which is presented to the visitor for identification and empathy (although curiously those who actually physically built the fortress are ignored). A particular element is the synagogue, a remnant of the role of Jewish traders in military supply. Bourtange's physical location on the Dutch-German border obviously endows it with a special role in defining and accentuating the separation of The Netherlands from Germany. There is a distinct feeling of defence of civilisation from the assaults of the barbarians beyond the frontier. The main dissonant elements in such interpretations is that Bourtange's foundation was an instrument of the Holland conquest of the resisting north which effectively created the existing national border which had not previously been a cultural divide. In addition its manning was frequently by mercenaries, most of whom would have been German.

Conclusion

These two cases, of the many possible fortress towns in The Netherlands and indeed Europe illustrate both the general problems but also variations from them. A major difference, important to many, concerns the fraught question of authenticity. At one level Naarden can be considered more authentic than Bourtange in that it was never demolished and that much of the physical fabric that exists today was first constructed in the seventeenth century. Bourtange was demolished and then subsequently reconstructed. In terms of the authenticity of the experience however the reverse could be argued. Bourtange presents a more complete and unambiguous fortress experience without the distracting elements of contemporary urban life. Naarden has a contemporary residential population with contemporary needs and facilities. Both towns demonstrate the need to extend the visitor experience, preferably to include higher spending and an overnight stay but both are constrained by the dominance of a single product and by the need to preserve the physical structures. It is difficult to introduce new tourism experiences other than those that relate directly to defence heritage and even to accommodate tourist support services. Both towns demonstrate an inherent paradox of heritage tourism, namely that the visitor is attracted by the structures of the commodified past but remains a creature of the present with modern transport, accommodation and other demands that cannot be satisfied without compromising the heritage experience itself. The shortness of the visit also necessitates the fortress experience being incorporated as just one element into a wider tourism package centred elsewhere.

At a deeper level the nature of the heritage product being bought and sold in fortresses needs consideration. If heritage is the commodification of a selected past to serve contemporary demands then what has been selected, by whom, and in the service of which needs, are important questions. At the most general level the interpretation of the artefacts of war present a number of fundamental difficulties. In essence fortresses were designed as the most efficient means of delivering metal balls or sharp fragments at high velocity into vulnerable human bodies with the purpose of disintegrating them. The fact that it rarely did so does not detract from that purpose nor facilitate its transformation into an entertainment product for a family leisure experience.

As well as strategies of interpretation there is a need to examine the content of the messages themselves and their contemporary relevance. The two simple obvious and related points are the 'defence of the realm' and 'germanophobia'. The story of Naarden and Bourtange (like much Dutch 'history of the fatherland' as it is officially titled and taught in the elementary schools) is a simple narrative of the successful struggle between good and evil. The peace-loving, civilised, protestant heroes defend besieged little Netherlands against catholic tyrannies (whether Spain, France or Münster). The stress upon the 'golden age' of the seventeenth century provides the best stories but the narrative and its message can be continued into the present, thus legitimating the liberal, bourgeois, capitalist, Protestant State founding mythology. The Dutch national identity problem stems not only from the need for cohesion around such a Holland consensus, but equally separation from the rest of the Germanic culture area. This whole national history approach, created largely in the nineteenth century, sits uneasily with both the ideological state structures of preceding centuries as well as the evolving pan-Europeanism of the late twentieth.

Finally national political legitimation is not the only contemporary use of this heritage. It is also marketed as a tourism product. The potential conflict between the political and economic uses is obvious but dissonance can also arise between tourism markets most notably here because the nearest potential foreign tourism market, especially for Bourtange, is German. There is nothing remarkable about the same historical resources being transformed into a variety of different heritage products intended for different heritage markets. It does however require accurate segmentation and sensitive targeting and the extent to which this is achieved here is perhaps the most important challenge in the selling of fortress towns.

References

Ashworth, G.J. (1991), *War and the City*, Routledge, London.

Bruce, D.M. (1999), *Urban Tourism for the Sustainable Development of European Towns and Cities: the prepare approach*, University of the West of England, Bristol.

NHTV (1993), *Bezoekersonderzoek: Naarden vanuit toeristisch perspectief*, Nationale Hogeschool voor Toerisme en Verkeer, Breda.

Scheltema-Vriesendorp, E.A.M. (1990/1991), 'De restauratie van de voormalige vesting Naarden 1964-1989', in *Jaarboek Stichting Menno van Coehoorn 1990/1991*, Stichting Menno van Coehoorn, Utrecht, pp.161-172.

Stichting Menno van Coehoorn (1962), *Atlas van historische vestingwerken in Nederland: Deel 3, De provinciën Utrecht en Noord-Holland*, Stichting Menno van Coehoorn, Leiden.

Theme 3: Heritage as a Strategic Policy Option

BENGT O.H. JOHANSSON

Modernisation and development are constantly producing those leftovers that we use to promote into 'Heritage'. In that sense heritage is often something that is unclaimed by agents of progress. The investors have left for other ventures leaving the constantly under-financed Conservation Sector to mourn in the dilapidated structures. As globalisation and restructuring moves on ever faster, more and more heritage is produced, designated and defended by this Conservation Sector. As no society can afford to preserve the actual amount of this recognised heritage, new uses and arguments have to be invented.

The last decades have witnessed many creative reuses of deserted structures. The nostalgia and curiosity regarding good old days have been exploited in various Disneyland-style installations but also and not without success in well-situated former industrial horror sites turned into 'theme parks'. Creative reuse of such structures for modern needs, like shopping malls, have sometimes helped communities to regain a lost vitality with or without the help of tourist industry. In Europe structural funds from different EU programmes have typically been exploited in such endeavours. Part of this strategy is to turn heritage into distinct place images that may help to regain a sense of pride among citizens in their environment. Such pride - so it is believed - is a necessary requisite for attracting business. A long-term goal is the establishment of more sustainable living environment.

Success stories like the constantly told turn-around of Glasgow seem to offer receipts for a fast track to new life for even the dullest place. As tourism is said to be the fastest growing industry in modern society, the first idea that seems to occur to those who want to save an ailing place is usually recreation facilities. The good thing with tourism is that it is a service industry and thus job intensive. One of the difficult things is that most tourism is local (we are not talking here about international sun-and-sea tourism) and must recruit customers from the neighbourhood, which thus needs to be fairly densely populated.

The studies that follow describe and analyse three different cases from the three participating countries: one industrial town after the dramatic closure of coal-mining in the UK, one former fortress town in The Netherlands, and one small ironworks relic in Sweden. Only the first one is situated in a densely populated area and may stand a chance of developing into a reasonably well-utilised tourist resort. All three examples could be described as 'Company Towns', where the company - or as in one case the garrison - has left.

The ancient coal-mining town of Bolsover in the UK, as described here by Black, is perhaps the most straight-forward case. Bolsover was included in the national campaign that followed the closing of coal-mines. It has thus received substantial economic support from various governmental funds in order to sustain. While some of the grants are used for the promotion of an ancient castle, most of the money is used to clear the area from industrial waste that is unattractive to visitors and inhabitants alike, regardless the historical value of these remains. This is of course a problem if you believe that an important rational for heritage preservation is the handing down of physical documents that have the power to transmit knowledge and understanding of the past. Forsvik represents the typical case of industrial archaeology with its group of production buildings hard to reuse for anything else than low technology purposes. However the option chosen was not so typical. According to its present director Lars Bergström, the strategy has been to build on the immaterial values of the former industrial community and find uses that are compatible with the identity of the place in a broader sense than just building preservation.

Nieuweschans in The Netherlands has not so far decided which way to choose in order to remain 'in business'. The *Schans* (fortress) is gone with its garrison but a substantial mass of heritage buildings remain, mostly in the form of houses. The options to be discussed, according to Ashworth, are several - one of them being the obvious choice of marketing the town as a frontier town of historical importance. The competition in heritage marketing is however severe, as no site is without history. In Nieuweschans as in most other places this calls for a prudent strategy that calculates with risks of failure. Ashworth's paper outlined such strategies that could be applied to heritage sites anywhere.

While the strategy in Nieuweschans is still to be resolved, Bolsover is already set on image improvement which can create interest in different kinds of investment, regardless of the town's own traditions but in the same time negligent to the matter of authenticity. On the other hand,

Forsvik in Sweden represents a quite different strategy: that of trying to reuse old industrial premises and equipment for new enterprises, which at the same time respect the integrity of the site and are able to produce commodities for a growing building conservation sector with means that take up the industrial traditions of the place.

This is a low-key strategy that certainly does not attract big capital but can work in small places with a limited work-force. Like Bolsover Forsvik is however still relying on external financing from the government and is thus risky. Its merits lay in the combined preservation of authentic work and authentic structures - in my mind a combination that should be attractive to the Conservation Sector.

11 Bolsover – after 'King Coal'

GRAHAM BLACK

> The severe blow that the closure of both Bolsover and Markham Collieries was to the local economy should not be under-estimated. This blow certainly knocked Bolsover down, but did not knock it out. There is a realisation that the town must pull itself up by its own boot-straps, a willingness amongst all concerned to pull together, to use whatever assets it has....to ensure that there is no further decline, and to nurture the viability, vitality and vibrancy of the town.
> (Bolsover Conservation Area Partnership Action Plan, 1996: 11.)

Bolsover is a small market town of some 13,000 inhabitants located in the county of Derbyshire, c. 5 miles east of Chesterfield and c. 25 miles north of Nottingham. It lies on the edge of a limestone escarpment which forms the east side of the Doe Lea valley. To the west, the face of the escarpment falls steeply downwards to the River Doe Lea. To the east, there is a gentle slope downwards towards Nottinghamshire. Its pasts lie all around it. At Bolsover, the escarpment edge is pierced by a steep-sided valley, with small exposed limestone crags on either side, known as the Hockley valley. The first Bolsover Castle was built on the limestone spur created by the Hockley valley in the late eleventh century and the remains of its successor stand prominently in the landscape, visible for many miles north and south.

Down in the Doe Lea valley, below the castle, is all the evidence of the once proud coal-mining industry and its offshoots. A late starter compared to the Durham, Yorkshire or Welsh coalfields, for most of the second half of the twentieth century the Derby and Nottinghamshire coalfield was amongst the most productive in the country, and its miners were amongst the best paid. In the 1960s Markham colliery, near Bolsover, was the largest single colliery complex in Western Europe, employing up to 3,000 people.

The situation changed in the early 1990s. What happened then, despite huge public outcry, was the sudden closure of pits across the UK which were deemed to have no long-term future, and the privatisation of the tiny remaining rump of the coal industry. Bolsover colliery finally closed in April 1993, following initial attempts at closure in August 1992: 450 miners lost their jobs. Markham colliery had closed the previous

October, with the loss of 1,200 jobs, although attempts to re-open it continued for more than a year afterwards. Until the closures, Bolsover's economy relied primarily on the deep-mined coal industry - not just the collieries themselves, but also the area headquarters of the National Coal Board, until it was closed in 1987.

As if this were not enough, the 1990s also witnessed the decimation of the local textile and clothing industry, faced with overwhelming competition from cheap imports. There was a close relationship between textiles and mining - the men worked down the pit and the women in the mills. Now both sides of the family were out of work. Unemployment reached up to 43% in some areas of the administrative district of Bolsover, which included the old town of Bolsover and surrounding villages, with average male unemployment running at over 16%. In addition, the shops in the centre of the town faced increasing competition for the reduced money available from the nearby regional centres of Sheffield and Meadowhall, and the sub-regional centres of Chesterfield and Mansfield. If Bolsover was to recover, it would require not just a dramatic economic restructuring but also a social healing and an environmental transformation - effectively, a new identity. From the outset, heritage was given an important role to play in this.

History

Unless otherwise stated, the following account is based on the *Bolsover Conservation Area Partnership Action Plan,* 1996. In Domesday Book, the Leuric of Belisoure appears to have been a modest sized settlement. In 1086, the manor was held for William Peverell, who probably built the first castle on the spur. Substantial earthworks from the outer bailey of this first castle survive, primarily within the Castle grounds. The castle passed to the Crown in 1155 and was substantially repaired and strengthened after a siege in 1216. It appears to have declined in importance through the fourteenth and fifteenth centuries and was eventually almost totally cleared in 1603 by the new owner, Sir Charles Cavendish, prior to the construction of the present structure.

Following this reconstruction, both Castle and town experienced a period of national prominence. King Charles I was entertained at the Castle three times. During the English Civil Wars (1642-1649) the Castle was initially held by Royalist forces, but was taken by the Parliamentarians in 1644 and eventually used as a prison. It was re-purchased by the Cavendish family after the Wars, but its use was gradually reduced and it ceased being a residence in 1883. It was eventually donated as a ruin to what was then the Ministry of Public Buildings and Works in 1945.

There is considerable evidence that Bolsover itself is a planned fortified town, laid out on a clear linear grid pattern with a central axis, (see Figure 11.1). It is thought that the section of town north of Town End and the Market Place was a later addition, as it takes a markedly different form to the original plan. Bolsover received its market charter in 1225-1226. In the post-medieval period, Bolsover became famous for the production of high quality buckles and spurs, but this industry appears to have died out by the mid nineteenth century. In 1890 the Bolsover Colliery Company Ltd sank the deep pit at Bolsover Colliery and in 1891 started construction of a 'model village' at New Bolsover, in the fields below the Castle. This was the first of a series of model mining villages developed by the Company and the first model mining village built on 'garden city' principles in the country:

> The Company have tried to make the lives of the workmen as pleasant as possible, and to give them such an interest in the place in which they live that they are content to spend their leisure in their own village.
> (Mr J.P. Houfton, Managing Director, writing in the Times, 1913.)

The model village remains largely intact and separate from adjoining development. The development of the colliery coincided with a major expansion of the town, from 3,662 residents in 1891 to 11,214 in 1911, almost the same as the present population. The nearby Markham Colliery opened in 1924 and the Coalite Smokeless Fuel Plant and Chemical Works in 1936. The town had its own Urban District Council until 1974 when it was amalgamated with adjoining rural districts to form Bolsover District Council.

Today, the main western approach to the town, from the direction of Chesterfield and the M1 motorway, is dominated by the magnificent terrace range of the Castle. There is much of heritage interest, in addition to the Castle and model mining village. All the older buildings, apart from the Castle, are of local magnesian limestone. The enclosed Market Place is dominated by a number of public houses of individual design. There are seven ancient monuments and 228 listed buildings, including all the red brick miners' cottages of the model village at New Bolsover.

152 The Construction of Built Heritage

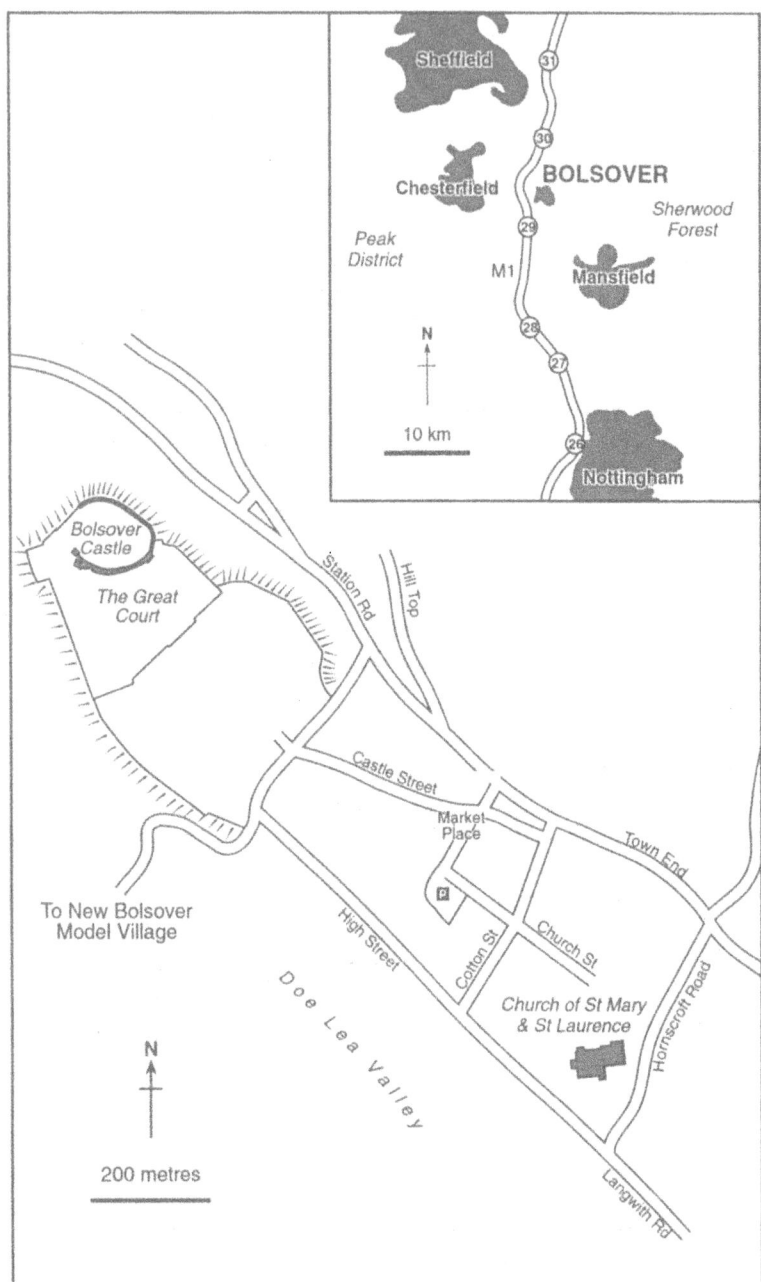

Figure 11.1 Bolsover town centre
Source: Reproduced from the Ordnance Survey based mapping by permission of Ordnance Survey on behalf of the Controller of Her Majesty's Stationery Office, © Crown Copyright ED 100017895

The centre of the old town and New Bolsover model village are both conservation areas. The Church of St Mary & St Laurence is largely thirteenth century, but may have an earlier foundation. It was extensively altered after a fire in the late nineteenth century. Overall, Bolsover has the feel of a small, stone-built market town heavily affected by its later industrialisation. The Castle itself is unseen from within the town until you come across Castle Street.

Regeneration

> The regeneration of a former coalfield area....must be based on a long-term strategy in which time and effort is given to assembling the full range of partners and stakeholders to secure their commitment to deliver innovative and integrated solutions that meet the needs of the people directly affected. To be effective, the process will require more flexible and pooled funding, packages of funding over a prolonged period, and more careful consideration of the integration of apparently disparate activities and of the timing of their implementation.
> (Interim Report assessing progress in the regeneration of former coalfield regions, published 21st September 2000.)

There was nothing new about the idea of colliery closures. They had always occurred as supplies were exhausted or geological problems made the workings uneconomic or dangerous. For decades before the 1990s, mines in Britain had been closing at the rate of 10-15 a year. What was different in 1992-1994 was the scale and immediacy of the closures. There was a huge public outcry in October 1992 when the planned closures were first announced, which forced the government to pay careful attention to rejuvenation. Equally, however, there was nothing new to the rapid decline and eventual collapse of traditional industries. Ports such as Liverpool, shipyards in various locations, the steelworks of Sheffield, the textile and clothing factories of Yorkshire - all had been through the process by the 1980s and had been forced to accept the need for fundamental economic structural change and to seek a means of achieving this.

Given the scale of public outcry over the 1992-1994 pit closures, and the seething controversy within the mining communities themselves, central government saw the priority as the creation of new jobs. There was no need to start from scratch. *British Coal Enterprise* (BCE) had been launched in 1984 to concentrate on supporting the economic regeneration of areas which had been affected by earlier pit closures, in the hope that new businesses would provide employment opportunities for those made redundant. In 1991, the European Commission had agreed a programme, entitled *RECHAR*, to promote economic, social and environmental renewal in former coalmining areas, with nearly half of the overall funding of, initially, £300m in grants and c. £500m in soft loans earmarked for the UK. Other potential grant aid was also available, including the EC's *European Regional Development Fund* (ERDF) and the UK's own regional grant programme, which was re-structured in 1994 as the *Single Regeneration Budget* (SRB).

Legal ownership of the former coalmine sites across England was transferred to Regional Development Agencies working alongside what became known as English Partnerships, the national agency for regeneration and economic development. Together they were given the task of reclaiming one of the largest portfolios of contaminated land in Western Europe - altogether the equivalent of c. 33 square kilometres. They were given a budget of, currently, £385m over 10 years for what was defined as the *National Coalfields Programme*. To this broad range of funds, could be added the small amounts of grant aid available from *English Heritage* and the much larger sums available for distribution by the various elements of the *National Lottery*.

English Partnerships (EP) have seen their role as being both a strategic and a practical one. The way forward has been to work in partnership with other agencies, from both the public and private sectors, and with local communities. They have also sought to maximise the amount of external resources focused on each area, by applying broadly-based 'funding cocktails' which allow a scale of resources to be brought to bear on an area much greater than could be provided by EP alone. This has disadvantages also, however, in the time spent seeking grant aid and in the uncertainty until the full package of funding for particular projects has been brought together - effectively a logistical nightmare. To put this in context, examination of the Bolsover economic strategy will reveal that it is, in terms of grant eligibility, within:

- Her Majesty's Government designated Assisted Area;
- East Midlands Regional Development Agency designated Rural Development Area;
- Coal Closure Area;
- North Derbyshire Derelict Land Rolling Programme Area;
- European Regional Development Fund Area (Objective 2);
- Single Regeneration Budget Areas 3, 5 and 6.

A key objective has been to diversify the economic base for the area. The District Council supported this by moving its main offices to Bolsover from Mansfield in 1994, bringing c. 150 jobs and their spending power to the town. A key thrust of the diversification has been the reclamation of former colliery sites and the establishment of business parks on them. Currently the Bolsover Business Park, situated on the old colliery site, has over approximately 12,500 square metres of accommodation, with 60 business units ranging from approximately 12-600 square metres. With a wide range of grant aid available, the park has attracted a range of businesses, now employing over 300 people - still fewer than once employed in Bolsover colliery alone, and with many of the jobs low paid, low skilled and part-time or seasonal.

The Role of Heritage

Past experience has shown that sustainable economic and social regeneration depends on much more than the provision of industrial units and grant aid. People must regain a sense of pride in their area. Skilled workers must want to live there and others to visit. Companies have a choice of where to establish themselves. Image and a sense of renewal is vital:

> It is against this background that many industrial towns in Britain, as elsewhere, have tried to replace negative images of dereliction, decline and 'smokestack' industries with images that stress change of function, modernity, and locational advantages which are often environmental in nature as well as economic.
> (Barke, M. & Harrop, K. 1994: 93-4)

The challenge is to change people's concept of a place - important in giving confidence back to local people as well as in attracting new people in. In the UK, the best known examples from the 1980s of re-imaging as a basis for economic regeneration were Liverpool, Bradford and Glasgow. Development was:

- 'property-led' - centred around the re-furbishment and up-grading of historic buildings and environments;
- supported by the establishment of major visitor facilities;
- backed up by the promotion of the locations to tourists, potential inward-investing businesses and potential residents; and

- targeted also at creating a centre from which visitors could explore the surrounding area.

The contribution that tourism, based on the heritage of the city and its environs, could make to employment was a major factor in the creation of local economic development strategies, even though many jobs were low-paid, part-time and seasonal. This view has not changed:

> The tourism industry in the UK is a key economic driver. It is worth some £64 billion a year (including day visits expenditure), around 7% of GDP. In 1999 1.8 million people were employed directly in tourism. It is recognised that tourism has created a quarter of all new jobs in the economy since 1990....Where traditional industries are in decline, tourism can play a key part in economic regeneration.
> (English Tourism Council 2000: 9)

All of this may be relevant to major locations but would it make any difference to somewhere like Bolsover?

The Bolsover Tourism Strategy: *Tourism's part in regenerating the rural coalfield* 1999-2004

Nationally UK government policy on tourism, in so far as it exists, attempts to view the industry not only as an economic driver but also in terms of sustainability (building conservation and environmental improvements), social development (leisure and cultural facilities, support for local shops etc.), economic regeneration (inward investment, employment, visitor spend etc.), and transport (improved access, improved public services etc.).

The UK tourism industry is highly fragmented, with many small operators. Much of the overall structure and strategy is provided by the public sector. Bolsover is part of the *Chesterfield Area Regeneration Team*, a partnership comprising Chesterfield Borough, North-East Derbyshire and Bolsover District Councils - an area which lies between the Peak District National Park (Europe's most visited National Park) to the west, and Sherwood Forest ('Robin Hood' country) to the east. While the area is rich in attractive countryside and heritage features, it lacks a strong identity of its own, with a low level of awareness amongst potential visitors.

Bolsover's Economic Development Strategy and local planning policy, developed after the collapse of the mining industry, looked towards:

- regeneration through diversification of the economic base;
- the encouragement of new employment-creating businesses and industry;
- the improvement of the social environment for residents;
- the conservation and enhancement of the heritage and natural environment.

Tourism was to be an important part of this. The North Derbyshire Coalfield was the subject of a major tourism study in 1991, with another for the adjacent area of north Nottinghamshire in 1994. Both pointed to the potential of the area in terms of its 'rural' resource and access - Bolsover is within 1.5 hours' drive of 20 million people, and within 3 hours drive of 30 million (the target 'short break' market).

Bolsover District Council recognised that it was in the early stages of potential tourism development and that it could not act alone. The challenge was to gain the active support and involvement of a wide range of bodies, from both the public and private sector. It was also appreciated that changes could not happen overnight. Central to the strategy was to recognise and develop the potential of the outstanding Bolsover Castle. To do so required a lead to be taken by *English Heritage*, the national body charged with providing expert advice on the man-made historic environment, but also directly responsible for those historic monuments, such as Bolsover Castle, which were in the direct care of the state.

The development of Bolsover Castle as a major attraction could be successful in its own right, but for it to have an impact on the town required visitors to explore beyond the Castle walls and car park. It was essential to establish a long-term programme of improvements to both the built environment of the town and the adjacent, scarred landscape. To do so would require action by the public and private sectors and community groups.

Bolsover Castle

> Four centuries on, love is still very much in the air at Bolsover Castle. Evidence of past passion is all around - from the erotic ceiling paintings in the Little Castle to the Romantic Venus Garden with its enclosed outdoor rooms and intimate love seats. Bolsover Castle exudes sensuality from every ancient yellow stone.
> (English Heritage 2000: 2)

English Heritage was established in 1984 as an agency independent of government. One of its key tasks is direct responsibility for the over 400 ancient monuments in State care. The creation of English Heritage - and even the agency's name - reflected a shift in policy away from a primary concern with the preservation and conservation of these monuments and towards one which placed much greater emphasis on the visitor experience. More recently, it has also explored ways of integrating its sites more effectively with local communities, as Pam Alexander, English Heritage Chief Executive put it 'I am keen to focus on our role in sustaining living, vibrant communities' (1997:2).

Bolsover Castle '...perched high over a motorway on the site of a medieval castle, in the midst of a former coal-mining area' (Guidebook) is a lot more than a ruin. It is one of the most remarkable buildings in England, and the 'jewel in the crown' of English Heritage monuments in the East Midlands. The impetus for a major programme of conservation work and re-interpretation came from English Heritage, but the funding of the work was only possible in the context of Bolsover's regeneration strategy. It was, in every sense, a co-operative project and followed on from widespread public consultation. The total cost of the project was made up from a 'funding cocktail' of:

English Heritage	£1,090,000
RECHAR	578,000
Single Regeneration Budget	416,000
ERDF (Objective 2)	728,500
Heritage Lottery Fund	188,000
Total	£3,000,500

To bring this funding package together involved separating the project into elements which linked to the grant-giving criteria of the different bodies. Thus, for example, the ERDF grant was linked largely to the development of the site as a visitor destination, and was therefore associated with job creation and income generation. By contrast, the grant from the Heritage Lottery Fund was used towards the conservation of a key part of the historic structure.

The Castle's reopening to the public in 1999 should be seen as both a flagship project for English Heritage and the tourism centrepiece of Bolsover's regeneration strategy. In its first business plan for the site, produced in 1997, the challenge English Heritage set itself was to increase visitor attendance from c. 35,000 before the refurbishment to, initially, c. 75,000. This target was exceeded in the first year following re-opening.

The business plan for the Castle reflected increased income from rising visitor numbers, the use of the site for activities such as the Bolsover Laser and Fireworks event and the potential of the site as a wedding and conference venue. The spreading of the income base is an essential element of planning for all heritage sites in the UK today, but also reflected the close relationship between English Heritage and Bolsover District Council.

Bolsover Conservation Area Partnership Scheme

The Castle was to be the focus for attracting visitors to Bolsover, but if they were to be persuaded to explore the town further - and if local pride was to be engendered - it was important to persuade owners to refurbish their properties, and to carry out environmental improvements both to the built fabric and to the surrounding landscape of the town. Conservation Area Partnership Schemes were created by English Heritage in the mid 1990s as a means of targeting the limited grant aid available to Conservation Areas which were in urgent need of assistance and where visible improvements could be readily achieved. They were to be a partnership between English Heritage and relevant local bodies and required financial and other commitments from both, over a three-year period. In Bolsover, the local bodies involved included the District of Bolsover, Derbyshire County Council, Old Bolsover Town Council, Bolsover Chamber of Trade, Bolsover Civic Society and Groundwork Creswell (a locally based environmental conservation charity funded from a variety of national, regional and local sources).

The Bolsover scheme was begun in 1996 and is now coming towards the end of its second three-year phase. Initial funding for work on the built environment ran at c. £50,000 per annum. This supported the restoration of a number of prominent buildings in the town centre and the production of a book on the history of Bolsover and of self-guided trail leaflets for the town centre and New Bolsover model village. In addition, English Partnerships funded the stabilisation of a major landslip which threatened a number of listed buildings on the High Street. The second phase has seen a greater financial input with works including the continued repair of targeted historic buildings, restoration work on the historic Market Place and environmental improvements along Castle Street/Middle Street, much improving the approach to the Castle and its links to the town centre. Cooperation between the District Council and the local Co-operative Society led to the construction of a new High Street store and three additional units, all in limestone, and a refurbished car park, at the same time as the Council was building its own offices. The work is ongoing and a third phase is anticipated.

The Carr Vale Millennium Project: Goundwork Creswell

Unless otherwise stated, this section is based on the report on this project by Groundwork Creswell (undated). In 1993, the District of Bolsover commissioned Groundwork Creswell to undertake a strategic study of the Carr Vale valley and the western approaches to Bolsover. The study generated a long term plan for ecological, economic and recreational regeneration of the valley, known as *The Carr Vale Strategy*. The Carr Vale Partnership of landowners, community groups and funding bodies was established in 1995 to drive a five-year regeneration programme for the area, with the work being managed by Groundwork Creswell. Starting with an initial allocation of £965,000 from the Millennium Commission, one of the bodies established to distribute UK National Lottery funding, an additional £2.7million of co-funding investment has since been realised by the partnership to help bring about the dramatic transformation of the post industrial landscape of the Doe Lea Valley.

A diverse range of inter-related schemes has been undertaken which has successfully addressed many aspects of the economic, social and environmental desolation resulting from the end of deep coal mining. Extensive areas of wetland habitat have been created from subsidence flashes along the river valley. The reclamation of colliery spoil has provided valuable public open space and areas of nature conservation with dramatic strategic viewing points. A comprehensive footpath and trails network has been created, not only to serve the local communities, but also to link with expanding regional tourism initiatives. The development of the wider access network has also involved a range of local environmental improvements around the villages of Carr Vale and New Bolsover, which have included new footbridges, allotment refurbishment, play facilities and the restoration of New Bolsover Village Green.

Significantly, the Carr Vale Project has directly involved the local community throughout the regeneration process. A programme of public consultation has been delivered alongside a wide range of participatory exercises, including design workshops, tree planting events and a variety of public arts projects leading to the siting of permanent community artworks.

Conclusion

In 1993 Bolsover was on its knees, with little reason to continue in existence, except perhaps as a residence for the retired and a dormitory for commuters. Instead, the local response was not to give in. The town has begun the process of re-discovering itself and has set in place a long-term strategy for both economic and social regeneration. From the early stages, it was realised that land and property-based development cannot on its own drive or deliver a sustainable revival for the community. A successful strategy must combine a range of agencies and activities and a cocktail of funds. All must progress together over a substantial period of time, and this requires a sustained vision. Heritage - both built and environmental - is a central element to this vision:

- A major visitor draw has been developed which provides a central focus. In its first year of operation, it exceeded its visitor target.
- The gradual refurbishment and reuse of fine historic buildings, the pedestrianisation of town centre streets and the reclamation of the local landscape is a continuing physical reminder of progress, essential to generating local pride and sense of achievement as well as in making the town more attractive to visitors.
- To date, the work has received the active support and commitment of local residents.

However, the problems continue and much remains to be done. Bolsover is still suffering job losses: 440 mainly female employees lost their jobs when the local Courtaulds textile factory closed in 2000. In February 2001 overall unemployment in central Bolsover stood at 7.0%, compared to 6.9% in February 2000. Male unemployment went down from 8.9 to 8.1, but female unemployment went up from 4.0 to 7.1 during the year. Unemployment in the other parts of Bolsover district stands at 6.2, 6.0, and 4.8%.

Perhaps as a reflection of a lack of local spending power, while shopping floor space increased from 5,540 to 5,854 sq. m. between 1994-98, empty shops remained around the same number (9 down to 8). Traders in the town centre believe the current situation in Bolsover is fragile. A recent extension of the pedestrian areas has resulted in a 10% trade loss for some, but the cause may be job losses resulting in lower spending rather than physical changes. Discussions continue to try to bring the market back to the market place, and to improve access for people with disabilities to the shops in the pedestrian area. A shop mobility scheme is to be introduced.

Even within a buoyant UK economy, it is difficult to maintain small market towns in the face of competition from larger shopping centres. The problems for Bolsover will continue for many years. Efforts to retain and slightly expand the shopping function will continue. The heritage challenge in this remains:

- To continue the programme of environmental improvements;
- To encourage visitors to the Castle to use the town and its shops;
- To promote the town and Castle more effectively as a visitor destination; and
- To assess the impact of work carried out to date.

References

Alexander, P. (1997), in *Heritage Today*, September 1997, English Heritage, London.
Barke, M. and Harrop, K. (1994), 'Selling the industrial town: identity, image and illusion', in J.R. Gold and S.W. Ward (eds) *Place Promotion*, Wiley, Chichester.
Bolsover District Council et al (1996), *Bolsover Conservation Area Partnership Action Plan,*, Bolsover District Council, Old Bolsover Town Council, Derbyshire County Council and English Heritage: unpublished report.
Bolsover District Council (1999), 'The Bolsover Tourism Strategy; Tourism's part in regenerating the rural coalfield', in *Bolsover District: Local Cultural Strategy 1999-2004*, Bolsover District Council, unpublished report.
DETR (2000), *Interim Report assessing progress in the regeneration of former coalfield regions*, Department of the Environment, Transport and the Regions, 21st September, 2000, London.
English Heritage (2000), *Attractions*, Issue 17, Summer 2000, English Heritage, London.
English Tourism Council (2000), *Action for Attractions*, ETC, London.
Groundwork Creswell (u/d), *The Carr Vale Millennium Project*, Groundwork Creswell, unpublished report.

12 Heritage in Economic Regeneration: the Case of Nieuweschans

G.J. ASHWORTH

Heritage and Small Towns

Amongst the many possible contemporary uses of the past, that of local economic regeneration has long been a prominent development option. Indeed in the North American experience, in contrast to that in Western Europe, the preservation of elements from the inherited built environment has frequently been little more than a by-product of an area regeneration powered by the search for new economic uses for local historicity (Ashworth and Tunbridge, 1990; 2000).

However, in most Western European cases including all three countries discussed in this book, the preservation of individual monumental buildings and the heritagisation of specific historical events and personalities largely preceded attention to the conservation and revitalisation of districts. Similarly, in the European case, the economic motive for heritage policies has traditionally played a role, generally subsequent and subservient to that of ideological legitimation and political identification. Nevertheless in the past decade there has been an increasing interest in the uses of the conserved past for contemporary economic purposes among both the conservation interest and place planners and managers. Specifically heritage has been viewed as a means of supporting ailing local economies. The reasons for this include the need for those involved in the preservation and care of monuments to find alternative sources of finance, and possibly also political justification, to support their ever expanding lists of conserved buildings and areas. Similarly those responsible for the management of places have become increasingly aware of the potential instrumental roles of heritage. Heritage can be used as an element in distinctive place images, promotable for economic goals. Secondly, it has a role as a contributor to a broadly defined local amenity of quality landscapes, itself an increasingly significant factor in the location decisions of enterprises and individuals. Thirdly it is a directly commodifiable resource for the production of local economic products (see Graham, et al., 2000).

Such general trends are apparent to varying degree and at various spatial scales in a number of the cases already reviewed. The focus here will be on the case of the small fortress town.

The Special Characteristics of Small Towns

It is difficult to generalise about the characteristics of small towns which can be as varied in their origins, functions and character as larger settlements. However, small towns have by definition a local economy that has a limited internal market and thus a high propensity to import from elsewhere. They are therefore particularly vulnerable to external trends over which they have little control and particularly dependent upon a few employment opportunities themselves vulnerable to the influence of external variables. Their options for development tend therefore to be restricted and their freedom of local action limited. On the other hand, relatively minor developments resulting in just a handful of jobs can have a major impact upon local economic vitality.

The special characteristics of fortress towns There are a number of contemporary characteristics of fortress towns in general that stem from their specific defence origins and which have been outlined in Chapter 10. Towns deliberately created as fortresses and sustained through many years by the paramountcy of the defence function, have a robust and compact physical endowment and strong historical associations. Additionally, however, they are frequently economically monofunctional, peripherally located and difficult to access. Fortress towns had few reasons to develop economic or social links with the surrounding countryside and military garrisons and their supporting populations were generally not drawn from the local region and tended to develop their own social relationships within the fortress. They thus tended to be inward looking, self-contained and self-reliant societies for whom the world beyond the walls was either irrelevant or indeed hostile (Ashworth, 1990). For these reasons defence towns at various scales throughout the world, of which the small fortress town is one category, have thus always presented particularly difficult problems of subsequent development (Riley, 1984).

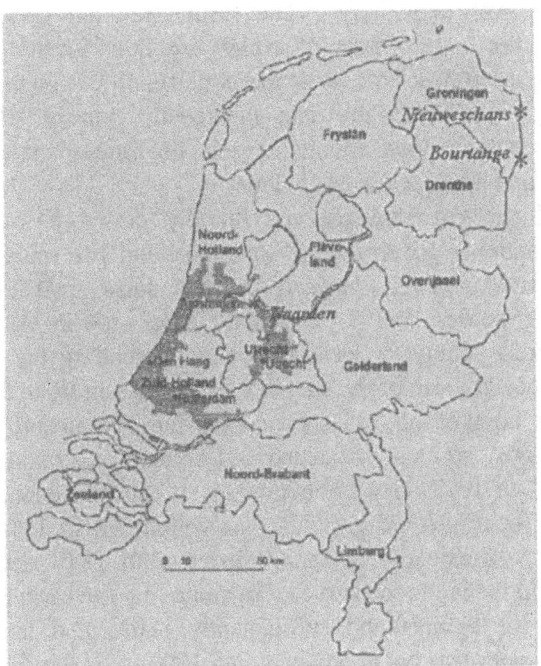

Figure 12.1 The location of Nieuweschans and other fortress towns discussed in Chapter 10
Source: G.J. Ashworth

The Problem of Nieuweschans

History

The political and military history of The Netherlands since its emergence from the Habsburg domains during the 80 years war (1572-1648) has been dominated by the necessity to defend the southern and eastern frontiers, from French and German attack respectively. An economic focus upon a westward orientated maritime trade and relative prosperity in the western provinces, encouraged the adoption of a military doctrine of static defence behind fixed fortifications on the landward frontiers. Investment in technologically advanced structures garrisoned by small, largely mercenary, artillery forces was an understandable option for a small country whose manpower resources and interests were dominantly invested in oceanic trade and commerce and in sustaining a naval hegemony. The result was the network of 'barrier fortresses' in the south (most of which are now located in Belgium or Northern France) and north-east of the country. Most of these remained in active military use until the late nineteenth century when they were superseded by defence lines, maintaining the doctrine of static defence, which were overrun and bypassed in May 1940.

The fortress at Nieuweschans (literally 'new fort') was established in 1628 by the national government ('States General') to replace an older and less well located fort at Oudeschans. Its subsequent history is easily summarised (Smedes, 1975). It had the responsibility of being the most northerly fortress along the eastern frontier, defending the route originally along the sea coast, north of the almost impassable morass to the south. It was updated on a number of occasions during the seventeenth and eighteenth centuries but only seriously besieged and taken by German invaders in the 'disaster year' of 1672. By the beginning of the nineteenth century it was clearly obsolescent and the garrison was withdrawn in 1815. The fortress however was not officially decommissioned until 1870 after which it was demolished and the land reconstructed by removing the earth banks and filling in the defensive ditches and subsequently reallocated largely to private owners. This work was not completed until 1907.

The town shared to some degree in the economic development, and specifically industrialisation, of northern Groningen province in the second half of the nineteenth century. The railway arrived from the city of Groningen in 1868 and the through route to the German network was available from 1874. The empoldering of the Dollart estuary to the north and the canalization of the Westerwoldsche Aa occurred in the course of the seventeenth century but the agricultural system of relatively large arable farms that developed, had little direct influence upon the town. The frontier location, and focus on frontier functions, did not encourage competition for the market and local service functions that are centred upon the towns of Winschoten and Leer, on their respective sides of the frontier. The only substantial industrialisation based, at least originally, upon agricultural processing was the straw board factory established in 1870, now 'Triton Karton' a division of Royal Dutch Paper (KNP), reprocessing waste paper products into cardboard.

The loss of the military function in the early nineteenth century was to an extent compensated by the continuation and even growth, as a result of increasing cross-border traffic, of the frontier function with Germany. Customs officials and border police (*Koninklijke Maréchaussee*) made up a high proportion of the total workforce. The steady decline in significance of the frontier culminating in the Schengen agreement (which includes both The Netherlands and the German Federal Republic) has led to the abolition of this function and indeed demolition of the buildings that housed it. Finally the EMU project and the elimination of national currencies in 2002, is even threatening the few remaining jobs in border currency exchange (GWK).

Population and Economy

The present population (1998) of around 1,500, in a total local authority (*gemeente*) population of only 7,000 is slowly declining with a negative natural and migration balance. With 800 inhabited dwellings the average size of household is just under 2. The employed population of the *gemeente* is only 2,800 (or 40% of the total) with a 22% unemployment rate in Nieuweschans itself. The largest commercial employer by far is 'Triton Karton' with 200 jobs, followed by the Fontana Hotel and spa complex with 55 and the transport company 'Gruppen' with 40. No other private employer has more than 10 employees (all statistics from Gemeente Reiderland, 1998).

Service provision in the town is limited to a single bank and a handful of local shops and cafés. These are not only sparse but lack a clear physical clustering, with the closest to a commercial centre being the *Buremaplein* area. Services of use to recreationists or tourists are also meagre. Neither of the two hotels cater specifically to tourists and there is a noticeable lack of visitor catering or shopping facilities especially in the historic core. Public transport comprises the hourly train service to Winschoten, the nearest substantial service centre and the city of Groningen; the rail service across the frontier to Leer has been discontinued and replaced by a three times daily bus service which has little relevance to Nieuweschans; there is no other scheduled bus service.

Definition of the Problem

A profile thus emerges of an ageing population, in which those over 65 (18%) outnumber those under 15 (15%). A relatively small workforce is dependent upon a handful of employers in a population with high unemployment and a high proportion of retired people. Related to this economic and demographic profile is a local society with a reputation for a certain introversion, if not passivity rather than enterprise. A tradition of dissent is reflected in the political composition of the local council (5 Labour party, 5 Communist Party and 3 Non-party council members).

The economic and social objectives of a development strategy are thus to retain existing employment levels, stem the outflow of young people to employment elsewhere by increasing local jobs, enhance the quality of the residential environment for existing and new residents, refurbish the local place image for internal and external promotion, improve local purchasing power and retain or expand the provision of local services.

Heritage and the Development Options for Nieuweschans

The Heritage Resources

The heritage resources available include a relatively large number (50+) of National Monuments (*Rijksmonumenten*) including the 'Guard House' and artillery officers quarters. These are spatially clustered in the centre of the town flanking the central exercise area, an elongated oval area, now planted as a park. This area was spared from intrusive through traffic in the past 20 years, probably fortuitously, and now forms one of the most unusual, attractive and charming high streets in The Netherlands. It is both the most visible reminder of the previous fortress existence and a major contributor to the urban feel of the settlement. Further afield, and more recently a number of so-called 'new monuments' designated under the 'Monument Inventarisation Project', of 1986 including amongst the industrial buildings a former grain silo on the Westerwoldse Aa and a semi-circular locomotive shed near the station. The 1961 Monument Act allowed the designation of so-called 'protected urban scenes' (*Beschermde stadsgezicht*) and the central area of Nieuweschans was so designated in 1974. Apart from the physical remains, the associations with the defence function and thus with nationally and even internationally significant events over almost 400 years, has left an exploitable legacy.

Visitor Facilities

There are few existing visitor facilities. There is a small fortress museum (and curiously and largely irrelevantly to the main theme, a sewing machine museum with limited opening hours). Facilities in the high street, the locus of visitor activity are especially limited with only a small tourism office, two small galleries/gift shops and the absence of any refreshment possibilities. There is therefore little to encourage a visit of more than a few minutes rather than hours, let alone days. Apart from the spa hotel, considered below, there are only two small hotels (26/36 beds).

The day excursion market is linked strongly to the main east-west route between Germany (Leer, Bremen, Oldenburg) and The Netherlands (Groningen/ Leeuwarden). Casual visits are encouraged by a Nieuweschans exit on the A7 motorway but discouraged by a confused and unattractive route into the settlement.

Two external factors could be encouraging. The first is the designation and promotion more than 10 years ago of the so-called 'Green Coast Road' (Bergsma, 1988) between Scandinavia and south-western Europe which runs through Nieuweschans. The idea was that holidaymakers, especially self-catering campers and caravanners, from Northern Europe driving to France or Iberia could be tempted to divert, visit and preferably overnight in the intervening communities. A more recent development is the plan for the so-called, 'Blue City', that is the creation of a high quality landscape through water inundation and tree planting between Nieuweschans and Winschoten. This would create new recreational opportunities and high amenity residential developments within easy reach of Nieuweschans and thus a potential day or holiday excursionist, market.

The Fortress Options

Perhaps the most obvious development strategy involves using in some way the unique and even dramatic heritage of the fortress itself. This is a resource almost too obvious and too potentially rich in its endowment, to ignore: but it is by no means so obvious how this can be done and to which end. There are four main benefits expected from the exercise of the fortress option in some way. These are an enhancement of the visitor attractions sufficient to encourage the development of commercial facilities for day, and ideally also staying visitors; some additional support from visitors for shopping and entertainment facilities serving principally a local market; an improvement in environmental quality indirectly beneficial to housing or commercial development; finally, and least quantifiable, a contribution to the image of the town whether for local or external consumption. The full range of possibilities can be reduced to two archetypes.

The 'maximalist' or 'Bourtange' option First, there is what can be labelled the 'Bourtange option' after the experience of the fortress town some 25 km to the south of Nieuweschans whose origin and historical development are similar to that of Nieuweschans in a number of respects and is outlined in Chapter 10. Reconstruction has firmly established Bourtange as a day excursion destination for both sides of the border. Bourtange was presented with stark economic imperatives a generation earlier than Nieuweschans and opted for a heritage tourism solution (Koeman-Poel, 1982).

Bourtagne can be regarded therefore in two contrasting ways. It may be a pioneering example demonstrating to other similar settlements what can be achieved especially in heritage tourism developments, on the basis of very meagre relict resources: although imitation might be discouraged by the question, 'how many 'Bourtanges' can a single regional market support?' Conversely it is used by some as a warning of the consequences of this line of development seen as a 'Disneyfication' of the past in which the settlement becomes no more than a monofunctional open air theme-park with the heritage tourism development effectively deterring the emergence of other functions.

The minimalist option   In contrast to the maximalist, or 'Bourtange', option would be the attempt to make some use of the historicity but falling well short of expensive, disruptive and, as far as monument conservation agencies are concerned, philosophically contentious reconstruction and developing some tourism activity but only as an ancillary to, and perhaps in combination with, other activities. This could comprise at the very least improved interpretive marking (which is currently almost non-existent), the production, marking and distribution of a 'town trail', and some physical symbolic representation of the fortress. The current single, curiously placed cannon at one end of the high street could be supplemented in various ways such as some symbolic marking of the site of the two gates at either end of the high street and more contentiously some attempt to recall the location of the bastions.

Other Development Options

The spa option   The discovery of curative water resources as recently as 1975, led to the building of spa facilities in 1985 (Projectgroep Nieuweschans, 1988), the first in The Netherlands; these were expanded in the following five years, to include extensive spa facilities including various forms of thalasso-therapy and an associated 134 bed modern hotel whose clientele are almost exclusively spa visitors. In practice the market for the spa, currently around 160,000 visitors a year, is dominantly German and it accords closely to the philosophy, and practice of health spas in Germany. Although the second largest employer in the town as well as the largest land-user, the spa complex and hotel is by its nature self-contained with few reasons why visitors should seek facilities outside. The links between the spa and the rest of town are limited. The possibility of combining spa and heritage day tourism is a development option.

The gateway option. The traditional official border function has now almost disappeared. The town remains however located on a cultural and historical frontier which remains a significant factor in much of the tourism and spa developments already mentioned. A specific means of exploiting this is the *Poort van Groningen* development being executed on the site of the former customs and immigration post on the A7 main access road. A tourist information centre, refreshment, souvenir and other retailing facilities will, it is hoped, exploit the characteristic of being 'first/last in The Netherlands'. Although an enterprising use of an otherwise derelict site and potentially offering some new employment, the main drawback of this development is its physical inaccessibility and lack of economic links to the rest of the town.

The residential/suburban option The stress in the above possibilities on new employment possibilities should not conceal the reality that a high and growing portion of the existing population is outside the labour market. Improvements in the quality of the residential environment are directly beneficial to much of the existing population but also raises the possibility of extending the residential function for those with work elsewhere. An attractive built environment to which heritage could make a major contribution, considerable investment in the recreational potential of the surrounding physical environment, together with relatively low property prices would seem to encourage an extension of housing for those with work elsewhere. This would maintain the local tax base, provide a market for local services and go someway towards rectifying the skewed demographic age structure.

Combinations and synergies The above inventory of the major options is of course neither comprehensive nor exclusive. Combinations of related activities, or activities making joint use of related attributes is not only possible, it is desirable. The 'Master Plan' that is in preparation (Gemeente Riederland, 1999) lays considerable stress upon integrating, physically but also perceptually, many of the different elements within the town discussed above. Specifically links are sought between the spa and its health tourists, the fortress, and its excursionists, the wider countryside recreation fostered by the neighbouring 'blue-city' waterside recreational development and lastly, the local residential function, revived by amenity improvements (Figure 12.2).

The means of shaping links between these elements are still under discussion but include transport routing realignments, some new access routes, the creation of attractive corridors between features for different transport modes and a range of broadly conceived promotional instruments.

The last includes many quite simple measures concerning public art, public space design and informational marking.

Some or many of these possible improvements to the quality and availability of the various products on offer could quite cheaply and easily improve the supply side of the equation. Successful combinations of products, or synergies in which the value added by such combination is greater than their sum, remain dependent upon a much more sensitive knowledge of the demand side. The users themselves, whether of the spa, the fortress heritage, the surrounding countryside features and the like, who will integrate elements to produce an individually customised combined product. Which experiences can or will be so combined remains uncertain.

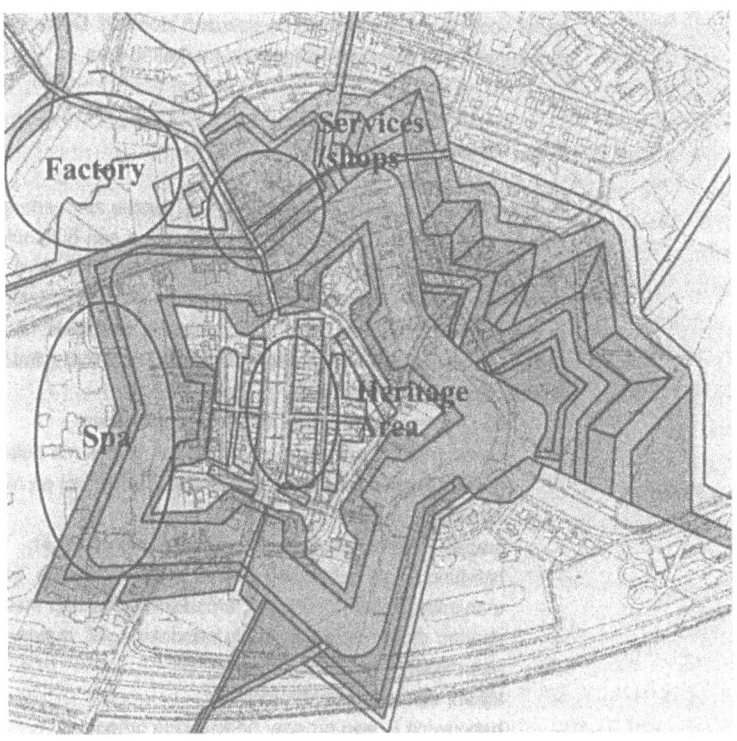

Figure 12.2 Modern Nieuweschans superimposed on previous fortress
Source: G.J. Ashworth

The Roles of Heritage in the Economic Regeneration of Places

While considering a single case of heritage in local economic development strategy, it is worth remembering that this example could be replicated by similar cases. All places have a past: all people have a culture. The very flexibility of heritage planning and of the wider use of culture in economic development, which allows almost anywhere this development possibility, also ensures that the competition will be intense. The cases of the use of heritage in urban economic regeneration policies which are described in the extensive literature (see Bianchini, 1993) can be misleading in so far as they are usually selected from among the successes. In a competitive field where failure is at least as likely as success, instructive lessons for new entrants in the arena can often best be drawn from the generally unrecorded failures. Certainly there is nothing automatic in the economic returns to heritage development. Demand for heritage products and experiences may be growing strongly in western societies but then so also is the supply of places capable of satisfying it (see the extensive discussion of the roles of heritage in economic development in Graham et alia 2000). The general conditions for the successful use of heritage in this way can be listed as the existence of an economic necessity, the development of heritage products that are competitive, the capability of discovering and serving an identified market and the existence of a local economy and society capable of maximising the benefits while minimising the costs. In all of these a large measure of good fortune in the location and timing of initiatives is probably also essential.

References

Ashworth, G.J. (1990), *War and the City*, Routledge, London.
Ashworth, G.J. and Tunbridge, J.E. (1990), *The Tourist-Historic City*, Belhaven, London.
Ashworth, G.J. and Tunbridge, J.E. (2000), *Retrospect and Prospect on the Tourist-historic City*, Elsevier, London.
Bergsma J.R. (1988), 'Planning of tourist routes: the Green Coast Road in the Northern Netherlands', in B. Goodall and G.J.Ashworth (eds) *Marketing in the Tourism Industry*, Croom Helm, Beckenham. pp. 89-100.
Bianchini F. (1993), 'The role of cultural policies', in F.Bianchini and M.Parkinson (eds), *Remaking European Cities*, Manchester University Press, Manchester.
Gemeente Reiderland (1998-9), *Gemeentegids*.
Gemeente Reiderland (1999), *Masterplan Nieuweschans; concept Bügel Hajema/Gemeente Reiderland*.
Graham, B.J., G.J.Ashworth and J.E.Tunbridge (2000), *A Geography of Heritage: power culture and economy*, Arnold, London.
Koeman-Poel, G.S. (1982), *Bourtange: schans in het moeras*, Stubeg, Hoogezand.
Projectgroep Nieuweschans (1988), *Hotel bij het kuurcentrum Nieuweschans: een markt verkenning*, Geopers, Groningen.

Riley, R.C. (1984), 'Military and naval land use as a determinant of urban development', in M.Bateman and R.C.Riley (eds), *The Geography of Defence*, Croom Helm, Beckenham pp. 52-81.

Smedes, J.J. (1975), *De Nieuwe of Langakkerschan*, Zaltbommel.

13 Forsvik's Bruk: A Tragic Industrial Closure or an Industrial Historical Success?

LARS BERGSTRÖM

Defining the Challenge

One of the major problems facing the heritage movement in Western Europe has been the preservation of, and the finding of new uses for, historic industrial premises following the widespread collapse of traditional industries. To date there have been two main routes forward. In a small number of cases, industrial failure has been followed by the creation of a historical museum or 'theme park'. More commonly, the shell of the buildings has been retained, while the interiors have been converted to new uses - as happened, for example, in the conversion to offices of Finlayson's factories in Tammerfors, or the building of a concert hall, exhibition halls and a conference centre in Norrköping's old industrial area, both in Scandinavia.

Is it possible to seek alternatives to these? In particular, can the application of a heritage management approach enable the revival of the original use of the site, redefining its period of closure as an unwelcome interlude rather than a final calamity? Can the heritage movement put into practice its belief that the retention of the values of the old industrial community is as essential to heritage preservation as the conservation of the buildings? Or is this a Utopian ideal, looking back to some sense of a more perfect past, but one that it is impossible to re-establish? Are there invisible and difficult to identify characteristics woven into the fabric of the community which cannot be planned for? How, for example, does one interpret the response of an old worker who, having lived for twenty years with the silence and dilapidation of a closed-down factory, once again experiences life and activity on the site and responds by demanding of the curator that 'Now the bell must get going!'

Forsvik's Bruk (*bruk* means works), at Lake Vättern in Sweden, is an industrial locality whose origins lie in the middle ages and whose engineering industry finally closed down in 1977. In many ways it was a typical example of the fate of traditional heavy industrial sites: as an atmosphere of decay and closure hung over the works, heritage interest was awakened by the importance of the surviving features. In the beginning, the problems were concrete ones; how to save the buildings from continuing dilapidation; how to finance their conservation? The justification for such action initially came from an investigation, which concluded that the settlement was one of the country's oldest industrial heritage sites. From the outset, there were visionary proposals for the future of the site, but also unanswered questions. Many of these remain to be solved.

Forsvik's Bruk Developments and Settlement Geography: A Short Historical Background

The settlement has the classic form, which it shares with the majority of older industrial settlements, that of being built around fast flowing water. In the case of Forsvik, by a waterfall between two lakes, which with its even flow was extremely attractive as a source of power.

The Pre-industrial Period

From the Middle Ages to the beginnings of industrialisation, the settlement consisted of a grain mill, a sawmill and a hammer mill. The saw mill, (water-powered) for the sawing of timber, and the hammer mill for the working of iron, are mentioned in Vadstena convent's register (*jordebok*) in the middle of the fifteenth century (Carshult, 1969; Julihn and Spade, 1982).
   The Convent used the waterfalls after a donation in 1410. The settlement had been a part of the large Forsvik's manor, which was under the same ownership as the industrial settlement from the mediaeval period until the 1930s. The saw mill and grain mill were, until the middle of the ninteenth century, mostly used for the manor's own needs, while the hammer mill, on the other hand, was developed during the late seventeenth century to become part of a large iron working complex. These production lines, and farming, were the major sources of income until the 1850s, when major changes occurred with the coming of modern industrialisation (Bergström, 1998; Julihn and Spade 1982).

The Göta Canal and the Foundation of Modern Industry

During the spring of 1810 the Göta Canal Company was given permission to start building a canal through Sweden, from Lake Vänern to the Baltic Sea. The digging and blasting started 1810 at Forsvik and Motala. An engineer from Scotland, Thomas Telford was engaged and with him a number of British craftsmen and instructors. To study British engineering methods and knowledge was invaluable for Swedish industry. The Canal was inaugurated in September 1832. The Göta Canal's influence on early industrialism was not just as an important transport route for goods, but also as a communications route promoting new technologies, the building of factories, the techniques of levelling and the production of maps and drawings (Gooch 1991, Bergström, 1998). For Forsvik's development, the building of the Göta Canal was, in many ways, the deciding factor.

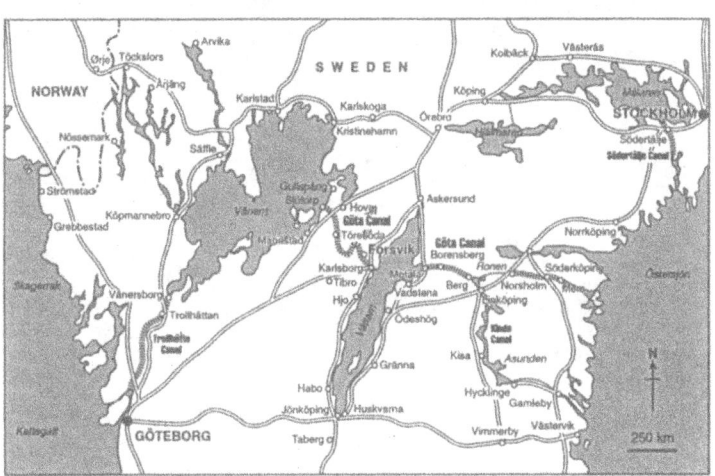

Figure 13.1 The Göta Canal still remains one of the largest construction projects carried out in Sweden. 190 kilometres long and with 58 locks. The importance as a transport route decreased in the beginning of the 20th Century. It is instead today one of Sweden's major touristic assets
Source:© AB Göta kanalbolag

The upswing for the settlement came soon after Forsvik's lock was finished in 1813. Partly new buildings were erected, like magazines and storehouses, the sawmill was made more effective in 1844, and more hammer mills were built. The modernisation of the settlement continued when bar iron production and ironware were abandoned in favour of engineering and casting in 1859, and a pulp mill was built in 1870. A new steam driven sawmill was built 1880 and the four hundred-year long period of a water driven sawmill came to an end. The expansion continued until the 1920s at which time the factories gave work to about 200 employees. Early on, as a consequence of the development of agricultural technique, various agricultural machines were produced. Technical innovation, and the growth of towns, led Forsvik to the production of pipes, pumps, valves, fire hydrants, lamp arms and lamp posts as well as cast iron components for the gas industry. Apart from these, Forsvik also produced everything from paraffin driven engines to cast iron sofas. Naturally, Forsvik's Bruk also did a great deal of work for Karlsborg's fortress and for the Göta Canal, which are both near by. During the 1920s a shipyard was built and the sawmill was developed and expanded along with the foundry to become the region's largest sawmill during the 1980s (Julihn and Spade,1982).

Figure 13.2 A view of Forsvik's Bruk 1930
Picture: Bo Bremsjö, Forsvik

The Industry's Decline and Fall

There are several reasons for the early technical knowledge. There was here an existing tradition of smithing and knowledge of the use of water power, but the transfer of technological knowledge from Great Britain meant a lot. Forsvik, however, never became one of the greatly expanding industrial areas. With the development of steam power during the second half of the nineteenth century, industry could be built near to the markets. The importance of the Göta Canal decreased. Railways and roadways became industry's transport ways, and this meant that Forsvik's Bruk was put more and more into the shade. While the factories in the towns were rebuilt and modernised, Forsvik kept its original character. From the end of the 1960s problems accumulated for the works.

During the 1970s, Göteborg's shipyards became Forsvik's main customer, which was the death stroke for Forsvik when the shipyards themselves died at the end of the decade. About 60 workers became unemployed and, as is usual when an industry closes in a small community, it was described as being a hard blow for the area. The sawmill, which had been separated from the ironworks in the 1930s, became more important and an increase in production reduced the problem, but the feeling that a new phase in the history of the community had begun was widespread amongst its 500 inhabitants (Julihn and Spade,1982).

The decline did not come all of at once. Many of the approximately 60 employees who lived in Forsvik tried to stay there by taking jobs in nearby places. Those who were near retirement age continued to live there, which meant that the population did not really decline until the 1990s. The total population decreased from 553 in 1983 to 414 in 2000 (municipal statistics 2001).

During the 1990s even the sawmill had problems. The number of employees was successively reduced. In 1999 the sawmill closed after having been in continuous existence for over 550 years and for the 39 employees there was no longer any work. In 1999 the school also closed and the shop closed a year later. During a period of 22 years, industry ceased to exist in Forsvik.

Table 13.1 Employees at Forsvik's Bruk
Source: Stiftelsen Forsvik's Industrial Heritage archives

Employees at Forsvik's Bruk	
1900	140
1925	200
1950	105
1975	85

Employees at Forsvik's Skogar	
1950	70
1985	235
1996	57
1999	39

Forsvik's Bruk becomes Forsvik's Industrial Heritage: The Terms of the Project

The situation in 1978 was that there was a dilapidated ironworks environment and a local authority, which did not know what to do with the problem apart from knowing that it was not economically capable of taking responsibility for the area. Threats overhung the buildings. The question of ownership was also a problem with many part owners. During the early 1980s, several attempts were made to run small scale industrial production in some of the workshops but all of these failed. There was neither a regional nor a local plan to attract any replacement industry (conversation in 2000 with Gunnar Ekfeldt, Municipal Commissioner 1983 - 1988). There was no local interest apart from feelings of shame over the state of decay and the only interest which was shown in the area came, at the beginning of the 1980s, mostly from heritage authorities.

Figure 13.3 Kalorifer – Air Hot Machine from 1877. The situation in 1983
Picture: Lars C Knutsen 1983

The local authority saw a support in this, and looked on it as a direction to lean towards. A working group was founded, with the Heritage Administration, the County Museum, the Municipality and local representatives from the sawmill and a local interest group. What speeded up the creation of the organisation was that money from job creation schemes could be used for conservation of buildings. The rebuilding started in March 1983 (conversation in 2000 with Lars C Knutsen, architect. See also Julihn and Spade, 1982).

This loose organisation, which however had very strong representatives, could begin the motley task of saving the buildings from a continuing decay. This large contribution during a short period of time made the working group see the seriousness of the situation and in 1987, the Trust for Forsvik's Industrial Heritage, was founded. Fundamental for future development, were sketches of ideas in 1985 for the use of the area as an industrial historical exhibition that should attract tourists. According to its statues the Trust shall 'take care of, manage and develop Forsvik's Industrial Heritage'. There were no provisions made for financing the Trust. Applications for grants were made afterwards. The shortcomings in both the governance and leadership of the site was clear, and it was agreed by the Trust's committee to look for money to finance a post in Forsvik. In 1992, a new conservation officer was appointed under the County Museum, with no more exact job description than 'to take care of, manage and develop Forsvik's Industrial Heritage'.

Forsvik's Bruk had thus been transformed from a living industry to an industrial heritage. Whereas 10 years earlier the pressure had been to modernise and rebuild to ensure continuing production, this had now been transformed into a priority to retain what was there. In this, the actors were united by the established principles of heritage conservation. The concept 'development' attracted, in this respect, diametrically different meanings from before. Apart from the County Administration Board having mentioned that this was of a national heritage interest for the cultural heritage management, strong visions of tourism had also been formulated (Idéprogram, 1985). The actors had however no clearly stated aims or expected results for the area corresponding to any economic plan. This, somewhat unclear, objective can be looked upon as a shortcoming but also as an opportunity. It gave free rein to alternative possibilities where one did not have to feel led by any fixed plan, which would be difficult to change.

The Actors and their Motives

The term actors are used to indicate those who initiated the project and are the institutional part of the Trust's committee. Actors who have joined at a later date, are described under the heading of 'Forsvik's Industrial Heritage becomes Forsvik's Bruk'.

The Representatives of Cultural Heritage

For the Regional Heritage Administration, Forsvik became a symbol for industrial history when the interest of the industrial cultural heritage was awakened during the 1970s, at the same time as the works finally closed down. In the light of the European Architectural Heritage Year in 1975 here was, well-arranged and gathered, to a large degree preserved, an environment whose buildings had housed eight different units of production. In addition, there were different categories of housing kept both for officials and workers. An industrial archaeologist had begun to establish a local authority industrial heritage inventory at the end of the 1970s, which meant focusing on Forsvik's Bruk as one of the most important cultural heritage sites to protect (Julihn and Spade, 1982). Forsvik ended up high on the prioritised list of economical contributions for preservation of buildings.

The Chief Conservation Officer and the County Counsellor was part of the Trust's committee until 1995 when the County Administration Board withdrew because of the risk of being an interested party in the handing out of grants. Today the role of Heritage Administration is to economically support the restorations.

Karlsborg's District Council The local authority saw two advantages in the conservation of Forsvik's Bruk as an industrial relic. The chance for increased tourism would be one of these, the other being the possibility of creating jobs. Furthermore, one supported a regional initiative that made an external financing possible instead of using the hard-pressed local government's money.

Forsvik's Skogar AB (Forsvik's Woods Company Ltd) When the ideas of an industrial heritage were put forward, Forsvik's Skogar AB was a healthy sawmill, which could boast of being Sweden's oldest sawmill. This fact was advertised in big letters on the buildings of the sawmill. When the old tube to the power plant was repaired with contributions from unemployment schemes, it was seen as an advantage to join the project.

The sawmill owned the south side of the stream where the power plant, the mill, the old pulp mill and the storage lay. Even if the company signed over the industrial buildings on 50-year leases, there was an economic advantage in getting the power plant going, which was apart from the actual project. Concurrently with the sawmill getting into economic difficulties, the power plant was sold to a larger power company.

The local ownership of Forsvik's Skogar AB ceased and the involvement in the Trust's committee was decreased. Soon after the bankruptcy of the saw mill concern in 1999, the sawmill concern left the board of directors of the Trust.

Forsvik's Industrial Heritage becomes once again Forsvik's Bruk?

The Different Phases of the Project

The first stage of the 'project of Forsvik' was a rescue phase. The big, leaking roofs had to be covered immediately. All the beautiful programmes of ideas about future use were soon drowned in the day-to-day problems of money, disputes of ownership and management. The allocation of money from government job creation schemes in combination with the Heritage Administration's money for building conservation for extra heritage costs, was the salvation for the buildings. A special problem was the fact that half the area was simply not available. It was owned by different diffuse companies, which had no great interest in what they owned and no wish to have any contact with the committee.

In a secondary phase, the council tourist bureau was actively working for the development of Forsvik as a place to visit. This meant that the planning for receiving visitors and arranging for staff rooms came up in the minutes. The responsible conservation officers were insistent that the work that was done correctly from a conservation point of view. Some of the suggested tourist-oriented projects were totally rejected. The new financial resources made available by the National Heritage Board, were used in building up a reception area in a factory building and an office for an in-the-future-to-be-employed conservation officer. The committee was unanimous in saying a management *in situ* was a must (Report of the Board in 1999 in the archives of Forsvik's Industrial Heritage).

The third phase can be summed up as the establishing of the Trust for Forsvik's Industrial Heritage as a part-regional industrial museum. The committee's role was changed from having been operative to becoming only decision making. The post of local conservation officer was soon changed to function as a director, submitting reports and executing decisions.

The eastern area of the site was taken over from a private company 1996. This company was just an owner with no industrial activity and now a new stage of building conservation could take place. The conservation was now led directly by Forsvik's Industrial Heritage, from applications for financing to prospecting, buying in and carrying out the work. The combination of economic support from the unemployment agency Heritage Administration and National Board Heritage made continued conservation work possible. This meant that after a while an organisation was built up with a line of different activities. The historical exhibitions were combined with theatre and art exhibitions. Students from the Univeristy of Göteborg wrote theses on both the intangible and tangible industrial environment.

In 1997 an engineering company moved into the newly restored office and the workshop. The company had earlier left the municipality during the mid 1980s but now saw the chance to return. A smith established himself at the same time as the reconstruction of a paddle steamer was started in the boat-yard buildings, where also a carpentry education for making models for cast-iron casting was started.

Figure 13.4 A view of Forsvik's Bruk 2001
Source: Stiftelsen Forsviks Industriminnen, Forsvik

The buildings begun to be reused and the tourism increased to about 30,000 visits a year. In 2000 there were over 37 man-years-work in the area and the situation had changed considerably from 1987, the year the Trust was started. Forsvik's Bruk had once again become an important and almost the only place of work in Forsvik.

Figure 13.5 Employees 1946
Picture: Sven Bremsjö

Figure 13.6 Employees in the area 2001
Picture: Oiva Isola

The Cultural Heritage Importance of Reconstructive Processes

The visions, which were worked up, were primarily aimed towards the public and other institutions to gain the interest for something that was reckoned to be valuable, but which was an unsafe adventure, both economically and politically. The area was in a bad state of disrepair and how could others be made to believe in that it would be possible to save it? There was also a vivid interest among young conservation officers at the County Museum and Heritage Administration who after the inventory discovered a new field of work which hitherto had not received so much attention. It is not possible to make a thorough evaluation yet. This is due in part to the fact that the project is still ongoing, which means that what is happening is not a project in the actual meaning of a restricted aim and a time-limited result, but an ongoing process, at present led from a cultural heritage perspective. That means that all the participants do not have to be included by this perspective. Also, more thorough research is needed about what has happened, and comparative studies to see if it is possible to find general traces.

We can conclude, going from the example of Forsvik, that a heritage conservation as a radical standpoint contributed to the works being preserved and even in such a way, becoming useful. No alternative for Forsvik's industrial area other than a total decay appeared. Karlsborg's District Council declined to buy the area in 1978 and there was no serious interest from any industrial company.

We can also see that the view on history and cultural preservation is changed with time and it is integrated into other fields of interest, thus creating problems, which give room for development. Let us take a few concrete examples. There is, within the area, a mass of tools and machinery which, from the conservation point of view, is to be looked upon as objects to be kept in its entirety for posterity. For the users, they are articles for everyday use, with the wear and tear that entails. The models from the model storage are marked in as museum objects but for the nearby foundry these are a resource, useful to 'borrow' while fulfilling certain orders. For the smith, the smithing hammers from the nineteenth century, are necessary in production and in the courses of education. In this, there seem to be a contradiction, but how should we function as an intermediary for our industrial history if these work tools were not to be used?

The bigger problems are the questions about administration for the future and historical representation. To start with the micro perspective Forsvik, the work with the works has meant that the concept of 'museological' felt passé. To reuse the name Forsvik's Bruk was considered more appropriate. 'Forsvik's Bruk' is a name which has lived for centuries and which shows/defines an identity which will live on into the future. In contrary to this the concept of 'museum' has a core of finality as though Forsvik's Bruk had come to an end.

References

Informants

Eva Björkman, Chief Conservation Officer, Västra Götalands län.
Gunnar Ekfeldt, Chairman of the District Council. Karlsborg.
Lars C Knutsen, Architect (SAR) Torekov.

Archive

Stiftelsen Forsviks Industriminnen (1987) Statutes.
Stiftelsen Forsviks Industriminnen (1987-1993) Minutes.
Skaraborgs länsmuseum (1995) Annual Report.

Andrén H., et al (1974), 'Forsviks Bruk - ett industrimiljöprojekt' *ArkitekturRumsplanering*, Publikation nr 2A, Chalmers Tekniska Högskola.
Bergström, L. (1998), 'Järnbron i Forsvik, exempel på teknikimport utan efterföljd' Stencil. Kung.Tekniska Högskolan Avd. för teknik- och vetenskapshistoria.
Bergström, L. (2000), 'Industrihantverk - värt att vårda?' *Nedslag i Västergötlands historia*. Skara.
Carshult, G. (1969), *Ett försök till kort historik över Forsviks egendom*. Forsvik.
Gooch, G. D. (1991), *Teknikimporten från Storbritanien 1825-1850. En studie av Göta kanals och Motala Verkstads betydelse som förindustriella teknikimportörer*. Linköping.
Isacsson, M. (1999), 'Det föränderliga industriarvet' i Eva Silvén and Maths Isacsson (eds.), *Industriarvet i samtiden*. Stockholm.
Julihn, E. (2001), 'Forsvik ett bruk i Sverige' (manuskript, byggnadsminnestext) Skövde.
Julihn, E. and Spade, B. (1982), *Industriminnen i Karlsborgs kommun*. Skövde.
Wikenros, I. (1997), *Forsvik och Göta kanal, Kanalbolagens arkiv berättar*. Mariestad.
Idéprogram till ny användning. Karlsborgs kommun Länsstyrelsen Landsting (1985).
'Sammanställning av idéer till användning av bruksområdet' (1984), Mariestad.
'Utredning kring Forsviks Industriminnen. Handlingsprogram för turistisk utveckling'. (1990), Karlsborg.

Theme 4: Heritage and the Restructuring of Symbolic Places

BENGT O.H. JOHANSSON

To discuss this topic is to discuss at the very pain threshold of professional heritage management. Much of contemporary discussion centres on the question of people's identification or non-identification with place-bound heritage. On a political level heritage is presumed to help cities or regions to flourish but not the least to help us to identify ourselves with the place where we live. This last argument is particularly favoured in programmes issued by the European Union and the Council of Europe in which the multi-facetted heritage of Europe should be brought to the consciousness of the inhabitants of all the European nations as a common heritage. The year 2000 was in fact proclaimed as the 'European Heritage Year'.

As Ashworth and McLernon/Griffiths directly or indirectly remind us by the discussions in their papers on Folkingestraat and Liverpool, the issue at stake can be boiled down to the matter of who 'owns' the right to interpret the heritage and enjoy it as such, in other words: who decides on its values and what it represents. Ashworth's point is that all heritage is customer created and may be sold to different consumer groups. The problem arises when in certain cases different interpretations of a site are dissonant. The memorial places of Holocaust are evoked as typical examples of this dilemma as the communities that once created the heritage in question were exterminated or are living elsewhere while the people who use the place today may be utterly alien to the history and traditions of those who once lived there. If heritage is taken as a kind of entertainment sold to tourists this certainly creates problems. On the other hand Liverpool is an example where the history and environmental heritage of an existing minority has been neglected but when elevated gives new interpretation to the environment. Monuments that were always 'sold' to tell the story of the city's economic success and as a kind of tourist entertainment are now, by descendents of an oppressed group, decoded to tell the brutal history of slave exploitation.

In the context of this book, which brings together contribution from the UK, The Netherlands, and Sweden, it may be useful to recall that in Sweden at least the term Cultural Heritage is a fairly recent import. Before that Sweden, in parallel to Germany, used the term *kulturminne*, equivalent to *Denkmal* in German (Cultural Monument). This concept was later broadened into Cultural Environment, a term still valid in the Swedish Act concerning monuments and ancient finds. While 'monument' stresses an object's capacity to communicate memories of any kind and pre-supposes an historic analysis of how, why, and by whom a certain object was created and what happened to it later on, allowing different layers of interpretation to exist simultaneously (here demonstrated in Landzelius' study of the Vänersborg museum), 'heritage' situates the object in a contemporary social context.

The words we use when discussing conservation are thus not neutral. Our use has political/ideological connotations. 'Heritage' tells us that we have received something from former generations to whom we are obliged and that those received things should be taken care of by us. 'Heritage' is thus a definition of persuasion with all the problems encountered in the following studies of Liverpool and Folkingestraat while to a certain extent a 'Monument', and more so 'Cultural Environment', indicates a use of that kind of historical source which might be called a remnant. The Australian National Committee of Monuments and Sites have coined the word 'Resource' in their so-called Burra Charter, stressing the cultural environment as a potential wealth. The good thing with that word is that heritage is looked upon as something that can be of use in contemporary life – another way of persuasion.

The entry on London's Parliament Square is about the opposite. It is about how a group of historic monuments are step by step transformed into not simply a cultural environment but into a sacred place, sacred to a nation's perception of itself. The story Burch tells us demonstrates how common heritage management practises – 'naming' and 'framing' – may produce and restructure a heritage site into something so secluded that it does rather not belong to us any longer – the paradox of the ultimate heritage being disinherited.

14 Liverpool and the Heritage of the Slave Trade

PAT McLERNON and SUE GRIFFITHS

This chapter is arranged in five sections. The first briefly outlines the built heritage of Liverpool; the second discusses Liverpool's involvement in the slave trade and the implications of this for the City's built heritage; the third outlines the crisis in race relations in the City in the 1980s and the significance of this for presentations of Liverpool's heritage; the fourth considers current heritage presentations of the City; the fifth offers a concluding evaluation.

The Built Heritage of Liverpool

Liverpool has a rich built heritage with over 1,000 listed buildings, eight of these listed at Grade I. In addition to its two twentieth century cathedrals (Anglican 1903, Roman Catholic 1967) the City boasts some fine eighteenth century buildings such as the Bluecoat Chambers (1717-1725) and the Town Hall (1749, reconstructed 1807), as well as a number of impressive Victorian civic buildings including St. George's Hall (1839), the Walker Art Gallery (1873-1877) and the Picton Library (1875-1879).

Albert Dock and the Albert Dock Traffic Office (1847) are among the eight Grade I listed buildings in Liverpool. Since the early 1980s Albert Dock and the adjacent Pier Head area have been a focus for urban regeneration, commercial development and tourism; this area now houses the Maritime Museum (opened 1984, the Transatlantic Slavery Gallery added 1994), the Northern Tate Gallery (opened 1988) and the Museum of Liverpool Life (opened 1993). In addition to these tourist attractions, the development along the Pier Head waterfront has incorporated twentieth century buildings such as the Mersey Docks and Harbour Board building (1907), the Liver building (1911) and the Cunard building (1916).

Among the Grade II listed buildings are many fine Georgian and Regency houses, foremost among them the houses built for wealthy merchants in Bold Street, Hope Street, Duke Street and Rodney Street. In the sixteenth and seventeenth centuries merchants had built their houses close to the docks, but as trading wealth increased they moved out to the higher ground. Duke Street, the first residential street to be built so far from the old town around the docks, was laid out in 1725 as a long avenue of fashionable residences; it housed wealthy merchants from among the most prominent families of the time. Hope Street followed in 1750 and Bold Street in 1780. In the Regency period stylish rows were built; Rodney Street was laid out in 1807 and the area around it developed in similar style. Clubs, concert halls, newsrooms and libraries were established for the use of this social elite, among those surviving being the Union Newsroom (1800) on Duke Street, and the Lyceum (1800-1802) on Bold Street.

Liverpool and the Heritage of the Slave Trade

The Slave Trade in Britain

The English first began to trade regularly with Africa during the sixteenth century. The Company of Royal Adventurers Trading to Africa was founded by the Duke of York in 1663 as a response to a growing demand in England for sugar. It was reformed in 1672 as the Royal African Company, a London based company which held the monopoly on trade with Africa, and for whom the exporting of slaves from West Africa to work in the plantations of the Windward Islands and Virginia was an unambiguous objective. The monopoly was ended in 1698 when Parliament opened the slave trade to merchants, and slavery was quickly established as a profitable commercial activity. Goods were shipped from the major British ports to Africa, where they were traded with African leaders for slaves, often the spoils of war. Slaves were then placed in cargo holds for the Middle Passage to the Americas. On arrival there the slaves were sold and goods such as sugar, tobacco and rice were purchased for the final voyage back to Britain.

The slave trade grew rapidly, bringing wealth and prosperity to the merchants and ports involved, generating new industries and employment, and producing huge profits. Between 1721 and 1730 British ships engaged from London, Bristol and Liverpool carried over 100,000 slaves to the Americas. By the 1760s Liverpool had overtaken Bristol and London and established itself as the pre-eminent British slave-trading port.

In the latter part of the eighteenth century a debate about abolition emerged, and after fierce and lengthy resistance from merchants the British trade in slaves was finally abolished in 1807. The anticipated collapse of trade links did not occur; links established with Africa, the Caribbean and the southern states of America endured, with continued and flourishing trade in goods such as sugar, tobacco, rice and cotton. Wealth acquired through slaving was profitably invested in continued trading along these routes, as well as in new links that were opening up with China and India.

Liverpool's Role in the Slave Trade

Liverpool's geographical position made it supremely well placed to develop as a trading centre, the growth of its mercantile shipping beginning with the Irish trade in the sixteenth century and expanding to the Americas and the West Indies by the late seventeenth century. In 1700 a Liverpool ship, the *Liverpool Merchant*, carried 220 slaves from Africa to Barbados, selling this human cargo for £4,239. From then onwards, slave trading formed a major element in Liverpool's overseas trade until it was abolished by Parliament in 1807. Historians of the slave trade have pointed to its crucial significance in the development of the city of Liverpool, for example Hugh Thomas in *The Slave Trade* states that 'the rise of Liverpool is a remarkable history, in which the slave trade played an important, perhaps even a decisive part' (1997: 246).

By 1740 Liverpool was sending thirty-three ships a year to Africa. By 1750 this had risen to seventy, and ships from Liverpool were carrying over half the slaves exported from Africa by Europeans. In 1771, 105 ships sailed from Liverpool to West Africa and carried to the West Indies 28,200 slaves. Trade dropped dramatically during the American War of Independence but revived when peace returned, and in 1784 Liverpool ships carried a total of 12,214 slaves. By 1798 the trade had grown to be more profitable than ever before, and almost 150 ships, the highest number ever, left Liverpool for Africa in that year.

The slave trade fuelled the expansion of local industries such as those in linen, glass, leather and metal goods; these produced goods for export and also equipment for the restraint of slaves, including chains, collars and shackles. Sugar refineries, rice mills and cotton mills grew up to process the imported goods for sale. The slave trade also gave a massive fillip to shipbuilding, and by the end of the eighteenth century the leading Liverpool shipbuilding firm Baker and Dawson had become the largest slavers in Britain. The ongoing programme of dock building throughout the eighteenth century also fostered the continual expansion of trade.

Few slaves ever set foot in Liverpool, but some did and were sold there. *Williamson's Liverpool Advertiser* for 1765 contains advertisements for slave sales, including the following entries:

> To be sold by auction at George's Coffee House betwixt the hours of six and eight o'clock, a very fine Negro girl about 8 years of age, very healthy.

> To be sold at the Exchange Coffee House in Water Street, this day the 12th inst. September, at one o'clock precisely, eleven Negroes imported per the Angola.
> (cited in F. O'Connor, *Liverpool Our City – Our Heritage*, 1990: 156)

The last public sale of a black slave to be recorded in England appears to have taken place in Liverpool in 1779.

The modern Pier Head road known as the Goree used to house the Goree warehouses, constructed in 1793 and destroyed by bombs in 1940. These were named after an island off the Ivory Coast where there was a prison for slaves awaiting transportation. It is not known whether any slaves were held or sold in the Goree, but the name does record and confirm the significance of slaving to the port.

Liverpool's Slave Merchants

During the eighteenth century, Liverpool's slave merchants were a dominant force among its social and political elite, prominent in the local government of the town and providing a number of mayors and Members of Parliament for Liverpool throughout the period up to abolition. Land owned by the town council and known as 'the wastes' was granted or leased out to its own members, who profited from developing it. Bryan Blundell, a former slave ship captain turned slave merchant, was Lord Mayor of Liverpool in 1721 and 1728. He erected the Bluecoat School (now Bluecoat Chambers) on land granted to him in 'the wastes', as a charity school for 'the training of poor boys in the principles of the Anglican church' (inscription on the façade). He apprenticed many of these boys to slave captains, or transported them to the New World to work as apprentices in the plantations. A map dated 1785 shows the new streets to be developed in the area around Rodney Street, and beyond this the land parcelled into sections by name, such as 'Sir Foster Cunliffe', 'Mr Hardman', 'Mr Blundell'. The names of these prominent Liverpool slave merchants are recorded in the names of its streets – Cunliffe Street, Hardman Street, Blundell Street and others.

Men like these used profits from their trading activities to finance building developments in Liverpool and establish banking enterprises to finance further ventures. Thus, the wealth gained from slave trading was a key feature in the commercial development of Liverpool, and commercial buildings openly displayed the source of this wealth:

> Liverpool was in no way shy about the benefits brought her by the slave trade. The façade of the Exchange carried reliefs of Africans' heads, with elephants, in a frieze.
> (Thomas, 1997: 246)

Although the majority of Liverpool's wealthiest merchants were engaged in the slave trade there was a small group who opposed it, including the campaigning abolitionists William Rathbone and William Roscoe. Their ships used the trading routes but did not carry slaves.

Liverpool after Abolition

The Liverpool Corporation and the majority of the town's merchants in African trade fiercely resisted all attempts at abolition for many years, claiming that the town's commercial prosperity would be ruined without the slave trade. Their power and influence were brought to bear in Parliament against abolitionist moves, and resistance was successful until 1807.

Following abolition, however, the anticipated collapse of Liverpool's trading pre-eminence did not happen; the existing trading links with Africa, the West Indies and the Americas continued to thrive. Africans were encouraged to produce goods for import into Europe, for example palm oil, which was used in soap-making and as a lubricant for industrial machinery. From the Caribbean and the southern American states, exports of sugar, tobacco and cotton found a buoyant market in Europe, many of the plantations that produced these goods continuing to use slave labour. Liverpool industries developed to refine these imported products, providing further investment opportunities for those with wealth derived from slave trading. Still more opportunities to invest this wealth were provided by the new trading routes with China and India.

The 1980s Crisis in Race Relations

The Problem Area of Liverpool 8

The Black population of Liverpool dates from the eighteenth century, the early residents being mainly servants of wealthy merchant families or seamen recruited from Africa, the West Indies or the United States to replace European crew who died or succumbed to disease.

A large proportion of Liverpool's current Black population is concentrated in the postal district known as Liverpool 8, which includes both the council estates of Toxteth and a significant number of Grade II listed buildings, among them houses built by and for Liverpool merchants engaged in the slave trade. In 1981, four years into the first Thatcher administration, the urban deprivation and poor race relations which had been rumbling ominously in clashes between the police and members of the Black communities in Britain suddenly ignited into major riots in the cities of London (the Brixton area), Bristol (the St. Paul's area) and Liverpool (the Toxteth area). The Thatcher government had in 1980 set up a number of Urban Development Corporations with a brief to achieve urban regeneration in cities where both the social and the material fabric were showing serious signs of degeneration. The 1981 riots increased the urgency of this programme in those major cities, and the Merseyside Development Corporation had the task of achieving visible physical regeneration in Liverpool. Its members decided on a strategy for urban renewal that was based on tourism and leisure attractions, the showpiece of their plan being the Pier Head and Albert Dock development. Albert Dock, which had fallen into dereliction in the 1970s after the closure of the South docks, was restored at a cost of about £30 million, and the museum-building programme around it was begun.

However, the Militant-controlled Labour Council in Liverpool opposed this government-led scheme and was determined to follow its own programme of urban regeneration, choosing to prioritise social and environmental programmes in the most deprived areas of the City at the expense of developing the city centre. The fabric of the city centre degenerated further, making it increasingly unattractive to business, retailers and the arts and entertainments sector. The Council's policies brought Liverpool to the verge of bankruptcy and precipitated confrontation with the government over local authority expenditure cuts, which the government was determined to enforce through harsh fiscal penalties. A series of municipal strikes followed, along with increasing social unrest, particularly in the racially sensitive parts of Liverpool 8. The political issues were eventually resolved with the disqualification of the Militant Councillors in 1987 and the imposition of a surcharge on them for fraud and mismanagement of public funds. The social issues remained to be addressed, in particular the experience of unequal treatment within the Black communities of Liverpool 8.

The Gifford Report: 'Loosen the Shackles'

A key response to this long-running disquiet within the Black communities in Liverpool was the setting up of an independent enquiry, headed by Lord Gifford and funded by Liverpool City Council. The main purpose of the enquiry was to enquire into police and community relations in Liverpool 8, and its preliminary report, *Loosen the Shackles*, was published in 1989, the title making conscious reference to Liverpool's slaving past. The recommendations of this report addressed the issue of what could be done to improve race relations in the City, particularly in the Toxteth area.

The Report began with a look back to the past:

> [T]he city's history shows an unbroken pattern of unequal and inferior treatment, ranging from outright cruelty and persecution to paternalistic contempt, through all the years from slavery to the present. The more that pattern is understood the easier it will be to put an end to it. (p.25)

This 'pattern' was understood and identified as an absence in Liverpool's representations of itself of any record of the part played by Black people in the history and development of Liverpool, and hence their effective exclusion from Liverpool's heritage and correlating marginalisation in Liverpool's present.

The Report went on to examine the presentations of Liverpool's past in its public institutions, in particular the Port of Liverpool exhibition in the Maritime Museum, recently opened in 1984. Focussing on Liverpool's slaving past, the Report commented:

> Modern Liverpool, while being aware of this shameful history, appears to try hard to gloss it over, if not forget it. The exhibition on the Port of Liverpool in the Maritime Museum on the Albert Dock has a panel which depicts the inhumanity of the slave ships, but the accompanying text glosses over Liverpool's role. (p.26)

It then cites this text:

> The slave trade did make a significant contribution to Liverpool's prosperity. However Liverpool's trading wealth was firmly established before it began to dominate the slave trade from the 1760s. (p.26)

The critical assessment of this exhibition in the Report provided substance for its first recommendation:

> that Liverpool's museums and public institutions, when they present Liverpool's history, should give a full and honest account of the involvement of Black people in the city, and that Black people should have access to their own history through an historical archive created by public funds. (p.26)

It is in the context of the Gifford Report that Liverpool's subsequent response to the heritage of slavery should be assessed. This impacts not only on presentations of the City's history, but also on presentations of its heritage of the built environment, since so much of this had its foundations in the wealth generated by the slave trade that fuelled the spectacular rise of Liverpool as a commercial and trading centre during the eighteenth century.

Whose Heritage? Current Presentations

The presentation of Liverpool's heritage is undertaken by the City Council and the National Museums and Galleries on Merseyside, a body of Trustees formed in 1986 to oversee all eight of the main galleries and museums on Merseyside. These bodies can influence the answer to the question 'Whose heritage?' through the ways in which they choose to present the history of the City's involvement with slavery and the slave trade.

'The Buildings of Liverpool' (1975)

The Buildings of Liverpool, produced by Liverpool City Council in 1975, exemplifies the approach from the period before the race riots and the Gifford Report. It covers most of the City's listed buildings and aims to encourage a greater awareness of the City's architectural heritage and to serve as a reference book. Its thirty-seven sections are each planned as a complete walk with an accompanying map. The authors give some initial contextual information, including one brief reference to Liverpool's slaving past: 'Liverpool merchants grew rich on the profits of the slave trade and the rapidly expanding commerce with the West Indies and North America' (LCC, 1975: 3). The rich architectural heritage is thereafter addressed in purely aesthetic terms, focussing on the architectural history and decorative detail of the buildings.

Since then, the challenges posed to the museum and heritage sector in Liverpool by issues around the slave trade and exposed in the Gifford Report have been acknowledged, and responses have been made. In 1990 the City Council replaced *The Buildings of Liverpool* with a new self-guiding booklet called *Liverpool Heritage Walk*. In 1994 the Maritime Museum opened a new Transatlantic Slavery Gallery housing a permanent exhibition with the title 'Against Human Dignity'. As part of its outreach work, the Museum has also adopted a Slavery Trail originally created independently in the 1970s by members of the Black community.

'Liverpool Heritage Walk' (1990)

This heritage trail for the listed buildings in the city centre is published as an A4 format booklet with a written commentary, illustrated by sketch maps and line drawings. It is sold throughout the City in stationery and newsagents' shops as well as at Tourist Information Centres. Reasonably priced, it offers good value for money for those seeking independent discovery of Liverpool.

The *Walk* presents a celebration of Liverpool's history through the built environment. The text does not ignore issues of slavery, but neither does it highlight them. Its forty-three A4 pages contain only three paragraphs on Liverpool's slaving past, incorporated in the account of the road called the Goree. The guide does not attempt to link other individuals, buildings or dates with this very brief sketch of Liverpool's involvement in slavery; neither does it make explicit the reliance of the architectural heritage on the wealth produced from slave trading and from investment and trade on the enduring routes after abolition. The apparent neutrality that it maintains on the issues around this wealth and its connections with slavery makes the text appear politically and morally neutral while privileging normative discourses and knowledge. In its own terms it is not controversial, yet it could be seen by those who may feel excluded by the neglect of Black history to offer an uneasy compromise with Liverpool's slaving heritage.

The Permanent Exhibition: 'Against Human Dignity' (1994)

The Transatlantic Slavery Gallery cost £535,000 and was funded by a grant from The Peter Moores Foundation, which had been looking for a home for an exhibition of the history of slavery since 1986. The Maritime Museum drew on many of it own archives and collections, and on those of the Liverpool Museum, in the final presentation. The focus of the exhibition, entitled 'Against Human Dignity', is not Liverpool; it begins in Africa and explores the diversity of African cultures, before moving to Europe and finally to Liverpool itself.

The exhibition was designed for a general audience, and to have national as well as local relevance. Museum staff were aware that interest from Black communities elsewhere in Britain might provide opportunities to widen their audience and increase the numbers of visits from those in ethnic minorities. At the same time, concern and suspicion had been expressed by some members of the local Black community about the Museum undertaking a project central to Black history, so interpretation was a sensitive and potentially dangerous area.

The choice of interpretive strategy seems to have been neutral and information-led, perhaps to avoid generating strong emotional reactions of guilt or anger in present-day visitors. The exhibition contains much text on interpretive panels, cased and interpreted objects presented in diorama, sound sticks which are used mainly for primary accounts, and a very low-key reconstruction of a slave ship representing the Middle Passage.

A permanent exhibition on slavery which gives voice to Black history and acknowledges its share in Liverpool's heritage could be read as a direct and positive response to the recommendations of the Gifford Report. However, the exhibition's anodyne approach and careful avoidance of controversial local issues point to what appears to be a failure of nerve: the result is offensive because of its neutrality. The exhibition renders the cultural significance of slavery as an aesthetic and erudite learning experience with little opportunity to express, let alone have, emotional responses. Outreach work is frequently undertaken by National Museums and Galleries on Merseyside, and efforts are made to ensure that collections and interpretive strategies in these presentations are socially inclusive. However, 'Against Human Dignity' is a permanent exhibition and thus remains static, a monument to its creators. The Trustees of the National Museums and Galleries on Merseyside remain in a potentially difficult tension between local and national requirements.

The Slavery Trail

A Slavery Trail of Liverpool city centre was researched and led by Eric Lynch, assisted by other members of the Black community, from the early 1970s. The Trail formed an element of his work as an anti-racial awareness activist in local Black politics, and later as a training leader for the City Council. In these capacities, Lynch was also involved in discussions during the planning and opening of the Transatlantic Slavery Gallery. Subsequently, the Museum's Outreach Officer offered training for the Slavery Trail guides, and publicity for it under the aegis of the Maritime Museum. This arrangement is currently in place. The Trail guides remain independent of the Maritime Museum, being paid directly by the visitors who join a specific tour. The guided Trail provides the opportunity for contact with a knowledgeable local resident who offers a personal interpretation of links between the built conserved environment and the cultural underpinning of the wealth and prosperity that gave rise to Liverpool's remarkable architectural heritage.

The Slavery Trail includes many of the City's listed buildings, but the guides interpret their architecture and decorative details in terms which relate to cultural meanings of the heritage of slavery rather than to the cultural significance of the architectural heritage. The Trail's approach may be illustrated by a comparison with the approach of the *Liverpool Heritage Walk*. A visitor following the *Walk* down Water Street is given the following information about one of its grand 1930s buildings:

> Barclays Bank, originally the head office of Martin's Bank, was completed in 1932 and was designed by the Liverpool architect, Herbert J. Rowse (1887-1963). It is his most lavish work. It was technically very advanced for its time with completely ducted pipes and cables and low temperature ceiling heating. The sculpture is by H. Tyson Smith (1880-1972) and the whole ensemble is one of the most impressive commercial buildings of its time. (pp. 30-31)

This is all the reader is told, and it concentrates entirely on the physical building. In contrast, the Slavery Trail guide draws the visitor's attention to the decorative panels in the bank's porch, showing slave children with bands round their ankles and wrists. The Trail then goes inside the Bank and visits the eighth-floor boardroom where the original carpet incorporates wooden sailing ships into its design. These visual symbols are interpreted as the Bank's own acknowledgement of the slave trade as the initial source of banking wealth in the City.

On the Pier Head, the *Liverpool Heritage Walk* presents the Port of Liverpool Building:

> Designed by Arnold Thornely and completed in 1907 it is built in Portland Stone. The building could be said to utilise the dome of St Paul's Cathedral rising above the centre of a Renaissance palace. The great majority of the docks system in the Mersey estuary is administered from this building. (p.23)

The Slavery Trail takes an alternative approach, based on the emphatic location of the building's meaning in Liverpool's relation to the slave trade. A wooden sailing ship over the archway, and the repeated dolphins and Neptunes in the decoration of the building, clearly testify to the City's maritime heritage, but Eric Lynch reminds the visitor that many of Liverpool's sailing ships carried slaves. He points out the chains on the gate pillars and reminds the visitor that such chains were a product made in Liverpool and used in slave trading. Lynch's awareness of his own heritage is incorporated into his personal narrative, along with anecdotes from his own life story and experience of racism, and accounts of his own research on which his interpretations are based. His narrative makes it clear that his specific purpose in leading the tour is as part of a wider racial awareness strategy. His stated aim is to stimulate his visitors to go and seek further information for themselves, as he did.

Conclusion: Sharing the Heritage of the Built Environment

The potential for heritage sector contributions to urban regeneration is a theme developed elsewhere in this book, and the general issues are relevant to the Liverpool experience. Regeneration of the Pier Head and Albert Dock area contributes to the tourist economy and serves as a flagship for cultural life in the City. The regeneration of Duke Street as a cultural quarter supports the Council's overall strategy. At the same time, with the museums and galleries switching to national funding, the significance of local accountability becomes an important issue. In Liverpool, this issue finds a focus in the official re-presentation of Liverpool's 'shameful history' (*Loosen the Shackles*, p. 25) and the onus on those who make the decisions about this to respond to the Gifford Report's recommendations about social inclusiveness.

As well as providing different strategies for interpreting the symbolic significance of the built environment, the three examples of interpretation discussed above exhibit differing attitudes to Liverpool's involvement with the slave trade. Some of the differences may result from institutional imperatives, professional decisions or responses to the local situation. Whatever the reasons, the result is that the official heritage presentations provided by the *Liverpool Heritage Walk* and the permanent exhibition 'Against Human Dignity' fall short of meeting the unambiguous recommendations in *Loosen the Shackles* that museums and public institutions presenting Liverpool's history should 'give a full and honest account of the involvement of Black people in the City'.

The *Liverpool Heritage Walk* offers few if any clues to the decoding of buildings and ornament in terms of different cultural emphases in the history and heritage of Liverpool, and thus misses opportunities to develop a heritage strategy for realising the present in terms of the past. The neglect of Black history and the absence of explicit links with the sources of mercantile wealth are serious omissions which inevitably raise questions about the ownership of the heritage that is thus presented. The architectural heritage is not compromised as architecture when the fullest possible information about the different meanings it may hold for different groups is acknowledged: the heritage value and significance can only be enriched by this approach. The current prioritising of architectural history reduces opportunities for reading the buildings through an approach based in cultural diversity where issues of Black history and the slave trade can be explored, and articulated towards anti-racial futures. The *Walk* is in need of revision, and those responsible for this should address the shortcomings in the next edition. One potential strategy would be for the markers and plaques around the City to be extended: further information relating to Black history and the slave trade could be added, and given as much public prominence. Failure to do this will prolong the commodification of the architectural history and contribute to the cultural neglect of significant social history, as highlighted in the Gifford report.

The Slavery Trail is a relatively new heritage presentation. It fills a noticeable gap and makes a significant contribution to reinterpreting Liverpool's past in a culturally inclusive way, but it occurs mainly at weekends, is dependent on volunteers and is available to very limited numbers of visitors. The current lack of any paper-based, self-guiding trail seriously restricts the opportunities for visitors to discover the alternative cultural meanings in Liverpool's historic architecture, and the shared Black and white heritage in Liverpool. Such a trail has been produced by the Bristol Museums and Art Gallery, in the form of an extremely accessible and well-presented booklet in A5 format called *Slave Trade Trail around central Bristol*. Visitors to Liverpool should be allowed a similar opportunity, and this is something that needs to be addressed by the outreach officers of the National Museums and Galleries on Merseyside.

However, the general context seems to veer towards convenient amnesia, the political forgetting and neglect of key issues in local and national memory. A major source of Liverpool's wealth is slavery; much of the money to build the fine houses listed as part of National Heritage came from slavery; some of these buildings were constructed by or for slave merchants, their families, their goods and their social life. The 'official' approach to Liverpool's heritage largely ignores and at best downplays this, and by doing so appears to deny links with slavery. This neglect amounts to erasure of the memory of slavery from the popular consciousness of Liverpool. The Gifford Report's recommendation 'that Black people should have access to their own history through an historical archive created by public funds' has not yet been implemented. In the City's streets and built environment Liverpool's slaving past is rendered invisible; there is no memorial to remind or inform anyone of this aspect of the City's heritage. William Roscoe (1753-1831), artist, art historian, lawyer, banker and ship owner, campaigned against the slave trade for many years, becoming an MP for Liverpool in 1806 in time to help push the abolition bill through Parliament. He is buried in Roscoe Gardens, a small and derelict old burial ground near Lime Street Station. Eric Lynch and his Slavery Trail helpers would like to see Roscoe Gardens adopted by the Council and made into a garden of remembrance for local anti-slavery campaigners and the countless victims of the Liverpool slave trade.

Meanwhile, the joint cultural heritage continues to be hijacked and interpreted as an exclusively white European aesthetic experience. One of the most recent additions to the publications celebrating Liverpool's architectural splendours is *Liverpool, City of Architecture*, written by Quentin Hughes, Professor of Architecture at the Royal University of Malta and currently Chairman of the Merseyside Civic Society. It is published by the Bluecoat Press, an independent local press attached to the Bluecoat Society of Arts which is housed in the Grade I listed Bluecoat Chambers; this press also publishes the *Liverpool Heritage Walk* for the City Council. A glossy 'coffee-table' book full of attractive, sun-drenched colour photographs, *Liverpool, City of Architecture* is a selective gazetteer of what art-historical traditionalists might judge to be Liverpool's prime architectural achievements. Its lengthy introduction and the effusive comments alongside each photograph make no mention of Liverpool's heritage debt to the slave trade. Instead, the City's built environment is located in an elitist aesthetic and an unproblematic haze of romanticised history. The Bluecoat Chambers itself is lauded as 'the ancient gem of

Liverpool' and its origin attributed to pious 'benefactors' who wished to give poor children a Christian education. The fact that this architectural gem was financed by the misery and degradation of the thousands of slaves transported on Bryan Blundell's ships and sold for his profit is entirely ignored.

References

Dresser, M., Jordan, C. and Taylor, D. (1998), *Slave Trade Trail around Central Bristol*, Bristol Museums and Art Gallery, Bristol.
Gifford, A. M. (Baron) (1989), *Loosen The Shackles: The First Report of the Liverpool 8 Enquiry into Race Relations in Liverpool*, Karia Press, London.
Hughes, Q. (1999), *Liverpool City of Architecture*, Bluecoat Press, Liverpool.
Liverpool City Planning Department (1975), *The Buildings of Liverpool*, Liverpool.
Liverpool City Planning Department (1990), *Liverpool Heritage Walk*, Bluecoat Press, Liverpool.
O'Connor, F. (1990), *Liverpool our City – our Heritage*, Bluecoat Press, Liverpool.
Thomas, H. (1997), *The Slave Trade*, Picador, London.

15 Layers of Meaning, Layers of Space: City Strolling and the Museum Gaze

MICHAEL LANDZELIUS

In 1834, the city of Vänersborg, in the Southwest of Sweden, burned to the ground. In the city centre, the church and the county governor's residence stood among the charred remains of wooden buildings. Almost sixty years later, in 1891, the doors of the Vänersborg Museum opened. Then, in 1994, the Vänersborg Museum once again opened up, but this time as a 'museum of museum history'.

Here, I attempt to conjoin these events in a critical interpretation of the Vänersborg Museum. I will situate the museum in its historical, spatial and political context, and suggest that an interpretation should consider not only the overarching role of colonialism, but also the impact of 'civilised' European bourgeois culture upon patterns of movement and behaviour in the city and the museum itself. The prefiguring notion is that the past should be conceived of as complex sites of tension-filled practices, taking place on different spatial scales entangled in one another (Lefebvre, 1991; Pred, 1990). The specific discussion of the museum building and its signifying function is rooted in a typological understanding of buildings which simultaneously takes social context into consideration (Markus, 1993; Pevsner, 1976). This combined approach enables an exploration of how the Vänersborg Museum accommodated a multitude of meanings in its differentiated impact upon urban practices of different groups and individuals.

The twofold ambition in this study then, is, on the one hand, to suggest an alternative to historicist accounts of the past that reproduce a mythology of temporal inevitability and progress, and on the other, to indicate creative possibilities of consciously politicised accounts in the presentation and interpretation of objects in areas such as conservation, museum studies and heritage management.

Patriarchal Markers on Display

After the Vänersborg fire of 1834, a new city plan was drawn up by Nils Ericson, major, canal-builder, and engineer at the neighbouring Trollhättan locks, together with the city surveyor and county governor, Paul Sandelhielm (Lindahl, 1965: 13-16; Paulsson, 1950: 56-58). The new city plan represented a conscious endeavour to tackle the fire hazard inherent in the buildings and spatial layout of the old timbered city. The new plan consisted of a regular grid structure of blocks and straight streets with the main motive and motif in terms of city planning consisting of a firebreak one full block in width and stretching four blocks from east to west.

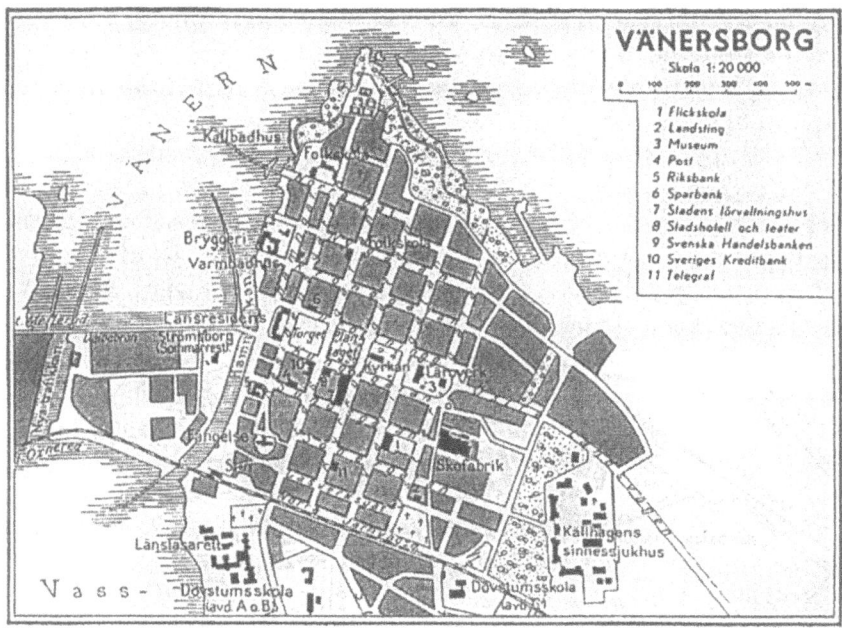

Figure 15.1 Map of Vänersborg The firebreak is located in the middle of the grid-plan, with the Vänersborg Museum in the centre of the map
Source: Svensk uppslagsbok, 1955

This firebreak created a 'front room' in the city, bounded by Queen Street (Drottninggatan) in the north, King Street (Kungsgatan) in the south, Residence Street (Residensgatan) in the west, and East Street (Östergatan) in the east. A market square occupied one block of the firebreak, but as a whole, the firebreak had a pronounced symbolic character, with the county governor's residence as a monumental closing motif in the west, the church in the centre surrounded by greenery, and space for other public buildings east of the church, where the grammar school and museum were to be erected. The symbolic ambitions were clear already in the naming of the three major streets delimiting the firebreak, but also in the local building ordinance. This ordinance was drawn up in 1835 and called for two-storey buildings along King Street and Queen Street (Hall, 1989: 44). In 1835, this did not represent a ban on larger and higher — on the contrary, it was a minimum requirement that was imposed in order to achieve a certain character of exclusiveness and prosperity. In this way, the new city plan immediately contributed to spatial segregation of activities and people.

The clear geographical structure of the plan offered possibilities to distinguish and create different regions in the city in terms of significance and meaning. The block-wide firebreak represented a *de facto* centre, yet the regular grid structure did not lend itself to making any particular point in the city plan appear to be more significant than any other. In this regard, the city plan was in line with the values propounded by the bourgeois revolutions in Europe: 'liberty, equality, fraternity'. In conjunction with Enlightenment-thinking, late eighteenth century neo-classicism in architecture preferred isolated, or 'liberated', buildings in space. In the Vänersborg plan of 1835, extant buildings after the fire were in this manner given full freedom to expose themselves in the firebreak. The buildings that represent power, prestige and respect either face towards or are located in the central firebreak, with the ecclesiastical and national institutions dominating. The county governor's residence and the church, which both survived the devastating fire were emphasised and placed as separate patriarchal markers in the city structure. All other buildings were subordinated to this structure and incorporated in the blocks — even the City Hall, the centre of the local politics of the burghers, was incorporated into the block structure.

The spatial order of the new city plan reflects a hierarchy of values that is neither self-evident nor necessary, but social and historical. With regard to Vänersborg, it has been stated that: 'The life of the main square is here not played out against a backdrop of functions to which it is immediately related — which is the case of a square dominated by bourgeois institutions — but in front of those façades that symbolise the nation' (Paulsson, 1950: 56). The spatial order that ensues from this layout make the citizens with their minor buildings appear to be subjects — the children of the state and the nation, over whom the supreme fathers keep their watchful eyes: God and the King, the priest and the county governor. Note that this city plan was drawn up before the abolition of the 'aristocratic' estate society and the introduction of democracy. In Sweden, it was not until 1919-1921 that reforms were introduced to give women and men without capital and property the right to vote and to make representatives for these groups eligible for election to political bodies.

Spatialising Surplus Time

The social dimensions of the new space of Vänersborg emerge with great clarity. The least appreciated citizens were exiled to the outskirts of the city. At the same time, the open main area in the centre of the city took on a more exclusive significance. Nothing was produced here. The function of the market square is the exchange and consumption of products, not production. The park area of the firebreak bears no relation to labour activity. It is therefore very clear that the central significance of King Street and Queen Street together with the park area are primarily linked with non-work, while a minor part, consisting of the market square, is devoted to consumption. The core of the city can thus be said to represent a surplus — a surplus of time to spend which is naturally associated with a surplus of economic resources. In the city centre, the surplus gains its collective spatial and material form: appropriated resources are turned into masonry as well as into leisurely walking practices.

A surplus of time is in a differentiating fashion linked with different people and strata in a population. However, space is also implicated, since the spaces that require a surplus of time and money to be utilised will inevitably be linked, in terms of the meanings associated with them, to the strata or people that can muster the needed time and money. In this way, the central firebreak gained its importance as a symbolic space of representation with more or less hidden rules of exclusion and inclusion, and this importance was manifested in the form of the promenade. The promenade was the most distinguished expression of this surplus of time, liberated from goals and utility it allowed expression of the will to show one's freedom from having to work in spatial terms. In recent years and relevant to the argument here, questions related to such expressively differentiating behaviour has been discussed with regard to the nineteenth century emergence of the *flâneur*, a figure characterised by an ostentatious visual consumption of his environment (Buck-Morss, 1989; Schorske,1981: 24-115; Shields, 1994; Wilson, 1995). Walking along King Street in the late nineteenth century, crossing the road at the grammar school and museum and then to stroll back on Queen Street towards the county governor's residence could in this perspective constitute a highly symbolic 'pedestrian uttering' (de Certeau, 1985) as part of practices that spatialised social distinctions.

Since the will to manifest the possibility of being free from work is part of a social and cultural pattern, the promenade is also part of the upper middle-class social life. Encounters and conversations in the street were a complement to tea and dinner invitations at home. The city's common spaces came to be dominated by one social stratum and became the external counterpart of the well-ordered drawing rooms in the upper middle-class home. In this fashion, the newly constructed urban space of Vänersborg participated in petrifying certain exclusionary and exclusive usages. The bourgeoisie displayed itself in the urban space in order to show its social position much like an object was displayed in the museum. In this main space of the tree-planted firebreak and its adjacent streets, the city's other residents are servants or passing visitors and onlookers. This is where the layering of spatial meanings and power begins to become evident.

Museum or Fancy Tombstone?

Johan Adolf Andersohn was 63 years old when, in October 1883 in a letter addressed 'To the Gentleman of the City Council in Wenersborg' he proposed the building of a museum. A prosperous grain merchant after many successful years in the business, he offered to cover the remainder of the cost himself if collection campaigns were unable to yield adequate funds. In his letter to the City Council, Andersohn proposed that the museum building should, as a first choice, be placed east of the church choir opposite the grammar school, probably with the main facade towards Rampart Street (Vallgatan), a cross street to the firebreak. Andersohn therefore requested that the city should 'make the proposed location available to the monumental building in question'. As a second choice, he was prepared to accept the 'rectory grounds at the eastern boundary'.

This suggests that Andersohn wanted to see the museum not only as a separate building in the firebreak, but also as close as possible to the very centre of the city. In terms of symbolic value of location, Andersohn found it appropriate to suggest behind the church choir, but in front of the large grammar school building. However, the City Council decided on another plot behind the grammar school, further to the east. This was right on the outskirts of the city, a place for horse-trading until the grammar school was built in 1869, close to the match factory, a fire hazard, and on ground that was still damp from the former moat. From Andersohn's point of view, it demeaned the museum to place it in the shadow of the much larger grammar school, further from the centre than he had wished and, moreover, on a site that immediately failed to command the respect of the strata of society whose values Andersohn shared.

From the point of view of city planning, however, when construction of the Museum began in 1885 on the plot decided upon by the City Council, the city's monumental firebreak was extended towards the east and a dignified approach to the city core was achieved also from this direction. And making its gables into backdrop motifs of East Street increased the museum's importance in the cityscape. The building was completed in 1886, but Andersohn did not live to see the Museum finished. When opened in March 1891, he had been dead for four years. Through these entangled factors, the Museum contributed not only to a more monumental appearance of the city, but also to the grafting of new meanings related to tombstone, mausoleum, and museum onto the significance of a site once allocated to horse-trading.

Typological Displacements

The Vänersborg Museum undoubtedly is a museum in the sense that it was built to house and display collections in order to achieve traditional museum-objectives in terms of education and enlightenment. However, understood in terms of building typology, is the Vänersborg Museum a true museum? With regard to architectural design as well as interior planning, it has been argued that the Vänersborg Museum is related to German nineteenth century museums by architects such as Gottfried Semper, Karl Friedrich Schinkel, Leopold von Klenze, and August Stüler, who in 1848 designed the Swedish National Museum of Art in Stockholm (Hall, 1989: 44-50). It has been argued that the Vänersborg museum on a smaller scale is a result of the same basic planning solutions as the National Museum by Stüler, which has made the building into 'a good example of the interplay between the architectural design and the underlying function' (Hall, 1989: 49). However, the same author has argued that while 'the National Museum is like many other museum buildings of the nineteenth century designed as a palace', the Vänersborg Museum in its outer and inner design is more suggestive of a bourgeois villa than a royal palace' (Hall, 1989: 46). I suggest that this confusion sparks an important typological question: a museum, a palace or a villa?

The Vänersborg Museum was designed by the architect August Krüger, who was born, educated and practised his trade in Germany before coming to Göteborg in 1852, where he had his office. Undoubtedly, he was familiar with German museums, and probably more so than with a museum in Stockholm. As an experienced architect, Krüger had a repertoire of solutions to choose from related to factors such as function, size and proper decorum of a building according to typological character. The typological question is: when he in the 1880s was contracted to design the Vänersborg Museum, a small museum in a countryside Swedish town, which template from his repertoire did he employ? The key issue is if the exterior and interior design follow the pattern and structure of the museum as a type.

Figure 15.2 Vänersborg Museum, main façade towards the west and firebreak. The façade is designed in neo-Renaissance style with the 1st floor — the *piano nobile* had it been a villa — articulated with visibly higher ceilings, arched windows and a more elaborate plasticity, particularly in the projecting section
Source: Regionmuseum Västra Götaland, Vänersborg

As institutionalised activity, the museum can be traced back to the private collections of Renaissance princes of the sixteenth century (Impey & MacGregor, 1985). It has been noted that '[t]he early researchers were generally private persons, without salary and offices. Amateurs, dilettantes and professionals were active within the same space. The determining factor for successful participation was above other things wealth and social prestige' (Nordbladh, 1993: 6). Thus, men of wealth brought objects home to their city palaces and displayed them as eye-catching parts of the interior in spaces for socialising and entertainment. Undeniably, such private collections were the most important predecessors of modern museums, and in this tradition, I suggest, is where we can situate wholesaler Andersohn, aspiring to being 'civilised' and 'cultured' by surrounding himself with exotic objects. However, with the Enlightenment, the idea emerged of museums as special institutions together with the question of the museum as a special type of building. This was a new conception in social as well as spatial terms, and it should be noted that the museum as it develops into a special type becomes, not only public and thus geared towards a new audience, but also a highly differentiated building with a spatial layering and zoning of distinct functions and accessibility (Markus, 1993: 171-212).

Figure 15.3 Villa Häbler, Dresden, street façade in neo-Renaissance style with the 1st floor, the *piano nobile*, articulated with visibly higher ceilings, marked out by arched windows and a pronounced plasticity particularly in the projecting section
Source: *Dresdner Architektur-Album: Bauten und Entwürfe*. Dresdner Architekten-Verein, Dresden o.J

During this period of development, the private palace style, with large undifferentiated rooms, was replaced by proposals for museums with new layouts that included several enclosed courtyards and a combination of large halls and adjacent galleries for exhibits, as well as libraries and extensive storage spaces and various types of workrooms. When museum building became widespread in the nineteenth century (Pevsner, 1976), and with growing collections, the spatial layering increased and the connections between the spaces in the museum changed. This happened in parallel with an epistemological shift from a descriptive phase where the exhibit coincided with the collection and the classification system, and thus with accumulated knowledge, to a phase where knowledge was seen as resulting from abstract reason and reflection. In museums of this period, we still find large halls as major exhibition spaces, but they were supplemented with smaller spaces (developed along the gallery system with niches) such as passages or corridors that allowed a practically infinite number of paths for the individual visitor to take.

When we look at the emergence of museum buildings in this typological perspective, it is doubtful whether the Vänersborg Museum follows the museum template. The building is much less like a museum than like another type of building, namely the villa with its roots in the Roman and subsequently the Renaissance country farm house. In the nineteenth century, the villa as a type is a distinguished free-standing residential building no longer linked with agriculture. When we look to Germany with this in mind we can see that many of the architects who were known for their museums were also well known as forerunners in villa architecture. The above-mentioned German architect Gottfried Semper, who was Director of the *Bauschule der Dresdner Kunstakademie* in the years 1834 to 1849 (Helas, 1985: 122), in particular had tremendous influence in large areas of Europe with his ideas of connections between architectonic form and function, often referred to as 'Semperism'.

In 1839, Semper designed the Villa Rosa in Dresden as a summer residence for a wealthy banker (Helas, 1991: 54). With this project, Semper transformed the villa into a comfortable but at the same time symbolically representative upper middle class city residence for distinguished residential areas. Villa Rosa was already an established model (Helas, 1985: 29, 43-51), known far beyond Dresden, when August Krüger, the architect of the Vänersborg Museum, began his German architect training. The striking similarities between these German villas and the Vänersborg Museum seem to indicate that the Museum, viewed as a building type, is a villa rather than a museum. Villa Häbler, designed in 1866 by the architect Karl Eberhard (Helas, 1991: 102-103), shows particular exterior affinities with the Vänersborg Museum, and is here chosen to illustrate my point.

The Museum as a Home Never Inhabited

If we compare the layout of the Vänersborg Museum with villas of the kind here mentioned, we can see how the spatial arrangements — of the entrance, of paths between foyer and rooms, and of connections between rooms — are generally based on the same overriding principles. The distinguished homes of the nineteenth century are characterised by ideals drawn both from French seventeenth century palace culture and from the Palladian resurrection of the Roman villa style. The atrium, which was drawn from the Roman villa, made up a central two-storey connecting room with stairs between the lower and upper foyers and usually also a skylight. In Sweden, the French ideals can be seen in the stately homes of the eighteenth century — a central large *salon* from which other rooms were symmetrically arranged as a long suite of rooms, an *enfilade*, with the gentleman's and the lady's rooms at the ends. During the nineteenth century, this type of spatial layout was modified but very much alive as an ideal which, when realised in its purest form, created an indoor promenade through the rooms that were parallel with the promenade for *flâneurs* made possible on the sidewalk outside.

We can also see that the space of the Vänersborg Museum is arranged in such a way as to combine the central atrium of the Roman villa with the French linear linking of the different rooms. Having disposed of utility functions to the basement and bottom floor, the upper storey of the German neo-Renaissance villas regularly extended the back-and-forth movement of the French linear plan into a possibility of an endless circular promenade. In the Vänersborg Museum, spaces of a practical nature and certain specific types of living spaces, such as kitchen and bedrooms, are not needed. The only residents are the museum caretaker and his family living in the cellar. When we study the plan for the upper floor of the Museum, we find the villa plan basically fully developed. The large rooms for socialising and entertainment in the private villa have simply been renamed in the public museum. Instead of a 'salon' or 'saal', we have 'ethnographic collections', etc. In the interior of the Museum, one can distinguish how renaming practices and the overlayering of different building types and spatial functions have created a complex signifying web.

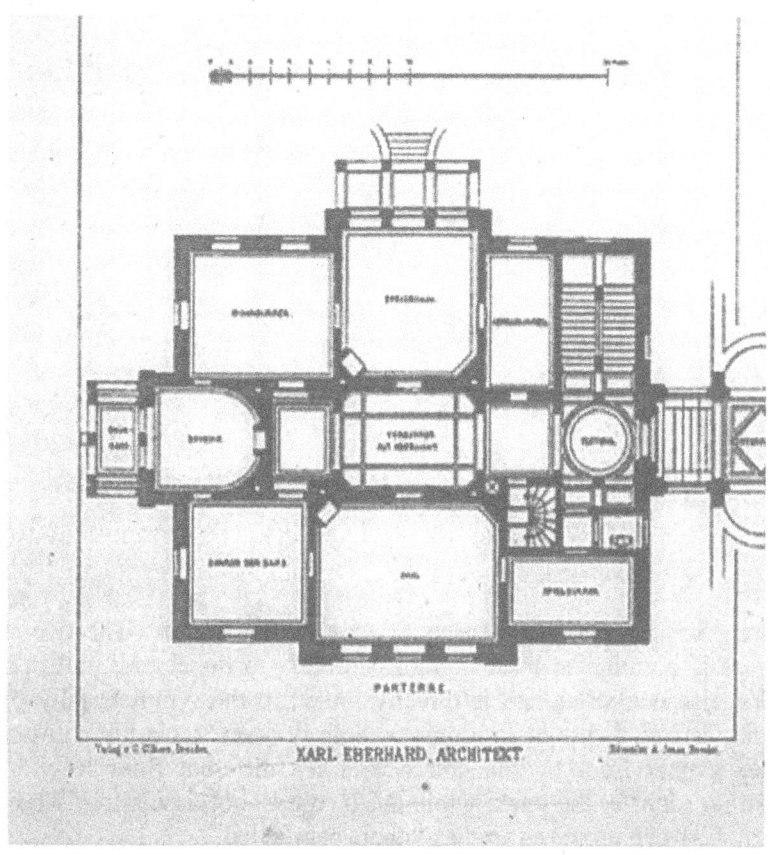

Figure 15.4 Villa Häbler, Dresden, ground floor plan In many Dresden villas, such as Semper's Villa Rosa and Villa Häbler by Eberhard the staircase was placed in one of the corners. This solution made access to corner rooms less convenient and excluded immediate contact between the staircase and the central hall as well as the *salon*. On both floors, circulation is possible directly between rooms and via the central hall
Source: *Dresdner Architektur-Album: Bauten und Entwürfe.* Dresdner Architekten-Verein, Dresden o.J

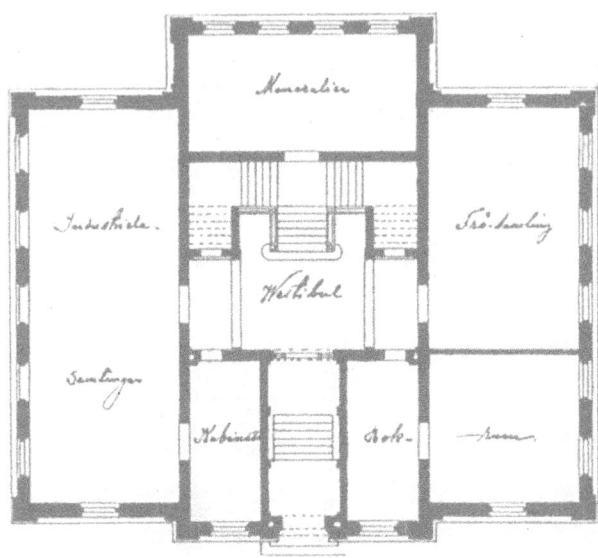

Figure 15.5 Vänersborg Museum, ground floor plan The two-storey staircase is illuminated by a skylight similarly to the central hall in many neo-Renaissance villas and is directly linked to the symmetrically placed entrance as well as the main rooms on both floors. Circulation is restrained on the ground floor by the entrance and a different floor level in the projecting room in the back, while on the 1st floor, circulation is possible directly between rooms as well as via the central hall
Source: Byggnadsnämndens arkiv, Vänersborg

In the Vänersborg Museum, you can go round and round and round again on a promenade that parallels the sequence of rooms for entertainment in an upper middle-class home. The spatial layering and differentiation which are a part of the museum building as a type are absent, as are the new types of rooms associated with epistemological shifts, and with the development of specific museum buildings. The spatial structure of the Vänersborg Museum resembles a view of knowledge in which classification, exhibition, and knowledge completely coincide, and corresponds to aristocratic and upper middle-class living habits with a representative social life in the centre. A distinguished city villa for the private collector is a congenial spatial and architectonic solution for wholesaler and donator Andersohn's collections. A home never inhabited, the museum *cum* villa *cum* tombstone is a free-standing spatial marker in the city fabric on a par with the buildings of the church and the state. As a counterpart to the county governor's residence in the west, the museum stands as a new patriarchal marker in the east, distinguishing Andersohn forever from the other citizens of the city. More equal than equals, a patriarch among fathers, wholesaler Andersohn built his private villa under the name of public museum.

The Unhomely House of Commons

In a public appeal for additional funding, Andersohn wrote that he wanted the Vänersborg Museum to become 'a sanctuary to which man and woman, young and old, rich and poor, like to make a pilgrimage in order to, under reflection upon what previous generations have achieved, gain a clear-sighted understanding of the width of their duties to those who will follow' (Andersohn quoted in Hall, 1989: 14). The museum is here conceived as a house of commons in a doubled sense: exhibiting tokens of types, individual specimens as representations of universal phenomena; but also allegedly a house in common for each and everyone in Vänersborg.

However, the notion foregrounding this article of the past as sites of tension-filled practices, suggests that individuals in different social positions applied very different frames of interpretation and thus approached the museum in radically different ways. For Andersohn, the objective of the museum was both educational and politico-ideological. His position was ambiguous, in that he, on the one hand, saw the museum as a means for the individual to develop 'his own greatness, his individuality and his independent personality' (Andersohn, 1883), while, on the other, he clearly saw it as an instrument of power. Thus, he rhetorically asked the City Council: 'Why would sovereigns, states, local administrations and individuals otherwise make such great sacrifices for this cause?'; as well as stressed the importance of museums in establishing 'true love of our mother country' (ibid.). His Eurocentric and racialised position becomes clearly visible when we juxtapose his statement that museums contribute to the fostering of a 'sense of beauty that manifests itself in more or less perfected artefacts of every cultured people [*kulturfolk*]', with his comment that 'primitive people [*naturmenniskan*] tattoo themselves and decorate themselves with shining bird-feathers' related to an 'innate but underdeveloped sense of beauty' (ibid). Andersohn travelled extensively in Europe, but had also visited Algeria, Egypt and Palestine, everywhere acquiring objects and registering customs of the people he approached. Letters from his travels were published in the county newspaper, and his reports witness that '[t]here is no doubt that he regarded European culture to be superior' (Hall, 1989: 13). While Andersohn's position on gender remains to be teased out, it is clear that the museum in his view is conceived of as both a racialised and a classed place for intellectual as well as ideological upbringing with immediate relevance for politics, governmentality and the execution of power. It has recently been observed that '[m]useums ... established exemplary models for 'reading' objects as traces, representations, reflections, or surrogates of individuals, groups, nations, and races and of their 'histories'', and that the museum as an institution thus has functioned as 'an instrument of historiographic practice; a civic instrument for *practising* history' (Preziosi, 1998: 509). The museum cannot be disentangled from the idea of bringing fragments from the world 'out there' into Europe, to be framed as orderly collections. In general, the museum produces 'an image of what would be visible from the very specific central-point perspective of a Europe masquerading as the present of the world's past; as the consequence of the natural evolution of civilization' (Preziosi, 1997: 5). In the particular case of the Vänersborg Museum, an odd microcosm was created in which collections of Egyptian antiques, china, stuffed birds from Africa, travel-souvenirs, ethnographical items from Russia, paintings, and local archaeological finds displayed a value-laden imagery of the larger world.

With regard to the Vänersborg Museum, this practising of history has to be understood in relation to the particular layering and nesting of different spatial scales and features. The placing of Europe as the centre of the world is repeated by the locating of the Museum in the heart of the city grid. Then there is the spatial correspondence between the movements facilitated by the central firebreak and the interior museum layout, as well as the resemblance between the spatial patterns of entertainment in a bourgeois home and patterns of museum visiting. We thus encounter a situation in which the practising of bourgeois social life as *flânerie* is duplicated as the practising of history as the *flânerie* of a European subject walking through the museum exhibits. The one is an endless promenade of presenting exclusive signs of European bourgeois wealth and manners, the other an endless suite of free-floating signs of otherness and others dressed in 'shining bird-feathers'. Hence, both the social and the cultural other was in an unhomely fashion simultaneously domesticated and excluded in this private villa *cum* museum. Understood as a kind of rhetorical device, the complex spatial layering and nesting of signifying elements that I have teased out here, could do nothing but confirm a self-affirmative mythology of never-ending progress, civilisation and modernity.

Concluding Remarks

Since 1994, the Vänersborg Museum displays four collections that originally formed part of the museum, and a reconstruction of the caretakers living quarters, furnished in 1950s style. This backs up a claim that the Museum is a 'museum of museum history' (cf. *Välkommen till Vänersborgs Museum*). However, I argue that the exhibited collections are simply simulations of themselves. The many critical interventions that a 'museum of museum history' could enable are squandered. From the beginning, the Vänersborg Museum was a schizoid conception, in which Andersohn paid homage to colonialism and racism as well as to democratic Enlightenment ideals. The stories possible to invoke through the Museum concern not only local conditions of social and cultural asymmetries, but also the injustices of the European colonial enterprise. In such a view, on display in any museum exhibit is also the [in]ability of curators to make careful choices and reconstruct forgotten, abjected and repressed meanings, as well as to produce new layers of meanings related both to the objects on display, and to the spaces in which they are displayed. The Vänersborg Museum could be engaged for the purpose, not only of showing four old exhibits, but also for critically narrating the deeply significant political history of seemingly innocent, and even benevolent, nineteenth century practices in a globally situated but local urban geography produced through tension-filled struggles between citizens of very different background, position, and prestige. In the present, the reconstruction of such forgotten stories would, for example, be important for the purpose of educating Swedish citizens about taken-for-granted patterns of racism and exclusion, in Sweden and in Fortress Europe.

Acknowledgements

I would like to thank Professor Bengt O.H. Johansson for his comments on an earlier version of this text. I would also like to express my gratitude to the helpful staff at Regionmuseum Västra Götaland, Stadsingenjörskontoret and Byggnadsnämndens arkiv in Vänersborg, Sweden, and at the Stadtarchiv, the Stadtmuseum, and the Stadtplanungsamt in Dresden, Germany. Finally I would like to thank STINT (the Swedish Foundation for International Cooperation in Research and Higher Education) for financial support.

References

Andersohn, J. A. (1883), 'Letter to the City Council' ('Till Herrar *Stadsfullmäktige i Wenersborg*') dated October 24, 1883, Vänersborg City Archive (Stadsfullmäktige protokoll i Wenersborg, 26 Oktober 1883).
Buck-Morss, S. (1989), *The Dialectics of Seeing*, The MIT Press, Cambridge, MA.
Certeau, Michel de. (1985), 'Practices of Space', in M. Blonsky (ed.), *On Signs*, pp. 122-45. The Johns Hopkins University Press, Baltimore, MD.
Hall, I. (1989) *Vänersborgs museum och Johan Adolf Andersohn*, Älvsborgs länsmuseum, Vänersborg.
Helas, V. (1985), *Architektur in Dresden 1800—1900*, Friedrich Vieweg & Sohn, Braunschweig.
—— (1991), *Villenarchitektur/Villa Architecture in Dresden*, Benedikt Taschen, Köln.
Impey, O. and MacGregor, A. (1985) (eds.) *The Origins of Museums. The Cabinet of Curiosities in Sixteenth and Seventeenth-Century Europe*, Clarendon Press, Oxford.
Lefebvre, H. (1991), *The Production of Space*, Blackwell, Oxford.
Lindahl, G. (1965), Karlstad 1865: 'Stadsbyggande för hundra år sedan', in M.Ståhl, N. Johnson and G. von Schoultz (eds.), *Värmland förr och nu 1965*, pp. 12-104. Värmlands fornminnes- och museiförening, Karlstad.
Markus, T. A. (1993), *Buildings and Power. Freedom and Control in the Origin of Modern Building Types*, Routledge, London.
Nordbladh, J. (1993), Untitled and unpubl. English manuscript; French version publ. 1993 as: 'Préface'. *Préhistoire Ariégeoise. Bulletin de la société préhistoire Ariège Pyrenéés*, vol. XLVIII, pp. 5-10.
Paulsson, G. (1950), *Svensk stad. Liv och stil i svenska städer under 1800-talet*, Bonniers, Stockholm.
Pevsner, N. (1976), *A History of Building Types*, Thames & Hudson, London.
Pred, A. (1990), *Making Histories and Constructing Human Geographies*, Westview Press, Boulder, CO.
Preziosi, D. (1997), 'Brain of the Earth's Body: Museums and the Fabrication of Modernity's Antiquity'. Paper presented in the forum: *The Future of Antiquity*, University of California Humanities Research Institute, Irvine, April 26, 1997.
—— (1998), 'The Art of Art History', in D. Preziosi (ed.) *The Art of Art History: A Critical Anthology*, pp. 507-25. Oxford University Press, Oxford.
Schorske, C. E. (1981), *Fin-de-siècle Vienna: Politics and Culture*, Vintage Books/Random House, New York, NY.
Shields, R. (1994), 'Fancy footwork: Walter Benjamin's notes on *flânerie*', in K. Tester (ed.) *The Flâneur*, pp. 61-80. Routledge, London.
Välkommen till Vänersborgs Museum. [Leaflet publ. by the County Museum, no year of publ. or author given.] Vänersborg: Älvsborgs Länsmuseum.
Wilson, E. (1995), 'The Invisible *Flâneur*', in S. Watson and K. Gibson (eds.) *Postmodern Cities and Spaces*, pp. 59-79. Blackwell, Oxford.

16 Shaping Symbolic Space: Parliament Square, London as a Sacred Site

STUART BURCH

The Process of Sacralization

The concept of a 'sacred sight/site', as defined in this study, exerts a strong and enduring influence on the policies, practices and outcomes associated with the construction of built heritage. In spite of linguistic and cultural differences notions of 'separateness, respect and rules of behaviour' are common to all sites that are considered to be sacred (Hubert, 1994: 11). Based on examples from the United States, Kenneth E. Foote has identified a general schema by which a place might become sanctified (Foote, 1997: 8-10). In the first instance it is clearly bounded and inscribed with a 'durable marker'; it is maintained over time and is likely to pass from private to public ownership; it becomes a place for ritual commemoration and a site for further monuments and memorials.

From this one can deduce that a heritage site becomes 'sacralized' by its ascribed associations' (Ashworth and Larkham, 1994: 19). The 'stages of sight sacralization' as set out by Dean MacCannell (1976: 43-48) form the theoretical basis for this chapter. Although this will be elaborated where appropriate throughout the text it is pertinent to mention at the outset the most salient aspects of MacCannell's argument. The process commences with two phases: that of *'naming'* succeeded by *'framing and elevation'* (original italics). These occur when a 'sight is marked off from similar objects as worthy of preservation' and circumscribed by 'an official boundary'. An additional level is termed *'enshrinement'* whereby the 'framing material' itself becomes characterised as sacred. The final two stages concern *'mechanical'* and *'social reproduction'* whereby the sacred sight is replicated and disseminated in the form of souvenirs and 'when groups, cities, and regions begin to name themselves after famous attractions' (MacCannell, 1976: 44-45).

This chapter will seek to apply a reading of the MacCannell thesis to a specific European city and on one particular locale within that metropolis: namely Parliament Square in the City of Westminster, London. The historical circumstances that shaped this urban landscape both physically and symbolically, as well as the various legislative regulations and cultural categories subsequently adhered to that site, manifest the naming, framing, elevating, enshrinement and reproduction that have led to its sacralization. In addition this chapter seeks to make an important modification to MacCannell's argument. It contends that, within the sphere of built heritage, the process of sacralization is less robust than MacCannell implies. The pattern is instead contingent and not necessarily linear or progressive, with sacred sites perpetually susceptible to profane incursions.

The Sacred and Profane

The association between memory and place is crucial to the establishment of a sacred site. Westminster is accordingly replete with historical connections: it has had royal and religious affiliations since at least the reign of Edward the Confessor. Indeed, the King's long-vanished Saxon palace 'directly gave rise to the present location' of the British Houses of Parliament (Factsheet 48). Adjacent to this, and built on the site of an eleventh-century Benedictine monastery, is Westminster Abbey, the nation's principal church. Westminster is therefore of great secular *and* religious significance. This serves to indicate a further important principle of what qualifies as a sacred site. The *Oxford English Dictionary* avers that the word 'sacred' has connotations with religion and worship. However, it also makes clear that it can equally refer more generally to something 'dedicated, set apart, [or] exclusively appropriated to some person or some special purpose' (Simpson & Weiner, 1989: 338-9, especially definition 2b). This non-religious quality is further enunciated in a series of 'special collocations': sacred artery, sacred vein, sacred axe, sacred circle, *sacred place* and so on (Simpson & Weiner, 1989: 339, definition 7). A sacred site therefore does not refer solely to a place of religion. In the case of Westminster the coalescence of church and state enriches it as a sacred site. The Houses of Parliament therefore garners as much approbation for being 'renowned world-wide as a symbol of democratic government' (Wheatley, 1997: 15) as Westminster Abbey does for being a place of religious pilgrimage. The image of the Houses of Parliament's Clock Tower is widely recognised and the chimes of 'Big Ben' are broadcast around the globe by the BBC World Service (Cannadine, 2000: 11), providing an instance of *'social reproduction'* whereby the nation and its capital city 'names' itself after its most illustrious attraction.

Discourse on the Abbey further compounds this relationship. The building was filled with sculpted memorials of national heroes throughout the eighteenth and nineteenth-centuries (Orbach, 1987: 214-6). A report in *The Times* newspaper of 1827 alluded to it as 'holding in its precincts the sacred ashes of the departed sages, heroes, patriots, and kings' (Anon, 1827). This description had been prompted by the funeral of the incumbent Prime Minister, George Canning (1770-1827). Following a suggestion that a monument be erected to the late lamented statesman within the Abbey, John Wilson Croker (1780-1857) declared with enthusiasm: 'Let his memorial be, as his remains were, placed in that sacred and immortal neighbourhood where are concentrated the most glorious names and recollections of our history' (Croker, 1828). Both quotations feature the word 'sacred' but in a markedly secular vein. Standing to the north of the Abbey– and of comparable importance in political as well as religious terms– is St. Margaret's Church. Established in the eleventh-century it later fostered close connections with parliament, as an entry in the *House of Commons Journal* of 1735 makes clear: 'It is as it were a National Church for the use of the House of Commons' (Wilding and Laundry, 1972: 663-664).

This is further evidence of the integration of the holy and the laical at Westminster. Parliament Square lies physically and symbolically at the juxtaposition of this crossing, located as it is to the west of the Houses of Parliament and to the north of St Margaret's Church and the Abbey. This seemingly 'empty' space was formed in the early nineteenth-century. From 1800 onwards legislation was enacted to enable the purchase and removal of a swathe of property in the vicinity of the Palace of Westminster in order to isolate the principal monuments: Westminster Abbey, Westminster Hall and St Margaret's Church (Crook and Port, 1973: 516). Throughout the nineteenth-century this area of Westminster saw the removal of a plethora of post-medieval accretions including small dwellings and workshops, public houses and coffee shops (PRO WORK 8/1A–8/10C). This allowed the grounds of St.

Margaret's Church to be enlarged and, 'for the sake of a better alignment', the area was 'cleared and levelled' and enclosed by railings (Parliamentary Papers, 1808, vol III:6).

This open space, grassed-over and planted with trees, became known as Garden Square, St. Margaret's Churchyard or Square, and by its present epithet: Parliament Square. This is analogous to the *'naming phase of sight sacralization'*. The differing titles for this site provide further evidence of the layering of religion and politics alluded to above. MacCannell opines that this initial stage of 'sacralization takes place when the sight is marked off from similar objects as worthy of preservation' (MacCannell, 1976: 44). Jane Hubert has similarly commented that, if

> something... is said to be sacred, whether it be an object or site (or person), [it] must be placed apart from everyday things or places, so that its special significance can be recognised, and rules regarding it obeyed. (Hubert, 1994: 11)

The clearance in the nineteenth-century of extraneous features from this centre of ecclesiastical and political power meant that the remaining structures were indeed situated at one remove from the commonplace and mundane in order that their 'special significance' could be appreciated. The stated reason for this undertaking was

> for improving the access and approaches to Westminster Hall and both Houses of Parliament... an accommodation much wanted upon all public solemnities: And to all travellers passing over Westminster Bridge, whether entering into or departing from the Metropolis, this clearance has at the same time opened a striking and magnificent view of Westminster Abbey in its whole extent, from Henry the Seventh's Chapel eastward, to the great Towers of its western entrance. (Parliamentary Papers, 1808, vol. III:7)

This conforms to the second phase of sight sacralization: that of *'framing and elevation'*. The former occurs when a site is circumscribed by 'an official boundary' whilst the latter 'is the putting on display of an object' (MacCannell, 1976: 44). It can be seen that this mode of separation and display occurred in Westminster whereby Parliament Square formed the 'frame' delineating the sacralized monuments. What is more, to perpetuate this 'magnificent view', it was directed that no subsequent structures be allowed to 'interfere with the view of the Abbey, from the intersecting centre of Bridge-street and Parliament-street' (in other words across Parliament Square) (Parliamentary Papers, 1810-1811,vol II:327). There are still in force today 'Strategic Views Corridors', consisting of cone-shaped areas three-hundred metres in width, which are intended to further preserve the aspect of the Houses of Parliament, St. Margaret's Church and Westminster Abbey, Figure 16.1.

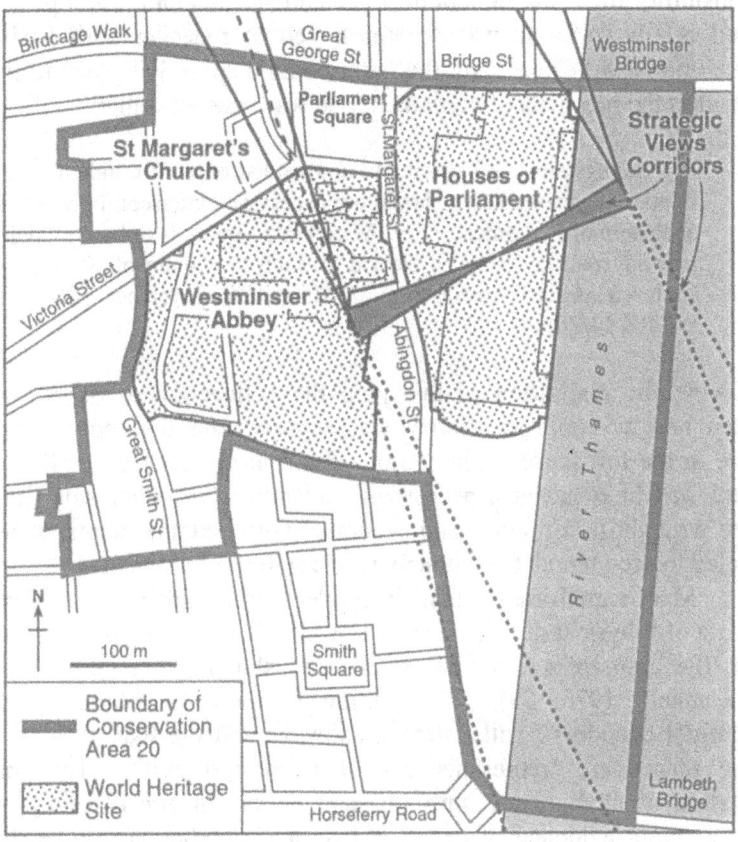

Figure 16.1 Parliament Square, London To show its relationship to the boundaries of the World Heritage Site and Conservation Area 20 and principal structures

Source: Reproduced from the Ordnance Survey based mapping by permission of Ordnance Survey on behalf of the Controller of Her Majesty's Stationery Office, © Crown Copyright ED 100017895

Barriers and Interstices

The ancient Palace of Westminster was almost entirely destroyed by a disastrous fire in 1834. Only the medieval Westminster Hall survived intact to be incorporated into the New Palace at Westminster designed by Sir Charles Barry (1795-1860) and A.W.N. Pugin (1812-1852). Following the death of his father, Edward Middleton Barry (1830-1880) assumed responsibility for the completion of parliament, including the open spaces around it. In 1864 he was commissioned to redesign both Parliament Square and New Palace Yard, the latter being a space lying just to the east and within the precincts of parliament. He was instructed that:

> The railings must be sufficiently high and strong to exclude a mob on important occasions, but should not necessarily interrupt the view. The enclosure of Parliament, or St. Margaret's Square is to be remodelled, and the roadway is to be carried through the centre of what is now enclosed (Alfred Austin to E.M. Barry, 26 November 1864; PRO WORK 11/20)

Following the realisation of this plan the connection to St. Margaret's Church and Westminster Abbey was severed and the space became a square in the full sense of the word. The central enclosure of Parliament Square was bisected by a pedestrian walkway. On either side, lined by ornate wrought iron railings and decorated with bedding plants, were sites intended to accommodate commemorative statues of eminent statesmen.

MacCannell argues that, in modern urban society, there exists a plethora of 'physical divisions' such as walls, fences, hedges and signs that mark 'the limits of a community, an establishment, or a person's space' (MacCannell, 1976: 39). In addition to these boundaries there are 'interstitial corridors': halls, streets, subways and the like. Within these public places are 'representations of good and evil'. The former– consisting of monuments, museums and parks– inspire respect. At the other extreme evidence of decay, refuse and dereliction invoke disgust. Taken in conjunction these two extremes 'provide a moral stability... that extends beyond immediate social relationships to the structure and organization of the total society' (MacCannell, 1976: 39).

The re-figured Parliament Square can be understood as an additional enhancement to the 'frame' of Westminster. Furthermore, it represents both a division *and* a corridor. It was intended as a permeable barrier: whilst facilitating the movement of people and traffic it was also a store for commemorative monuments intended to both remind and inspire. Although the railings of New Palace Yard were similarly intended 'to exclude a mob' they were not to 'interrupt the view'. In other words they were to serve as barriers to any unwelcome elements of society whilst enabling those willing to engage in the ritual of good citizenship to admire the nation's political and religious shrines.

Parliament Square's retention of memorials of the past is complimented by its facilitation of the orderly transportation of people and business in the present: it stores and regulates. In the 1850s the journalist George Augustus Sala (1828-1895) wrote that Leicester Square functioned as 'the liver of London' (Sala, 1872: pp174-175). The liver acquires the products of digestion, breaks down fats, produces bile and blood-clotting factors and expels toxins such as alcohol from the blood: it therefore serves to store and regulate (Dresner, 1995: 538). In contrast, Sala described Westminster before the urban clearances as 'a *cloaca* of narrow, tortuous, shabby, stifling, and malodorous streets' (Sala, 1894: 78). A 'cloaca' can be defined as 'a sewer; a cavity in birds and reptiles, in which the intestinal and urinary ducts terminate' (Schwarz, 1997: 193). This represents the two extremes in the 'interstitial corridors' of modern society: the one before the process of sight sacralization had begun and the other following the phases of 'framing and elevation'.

Re-framing Sacred Sites

The delineation and ornamentation of Parliament Square in the late 1860s meant that it had (to reiterate Kenneth Foote) itself become a place for ritual commemoration and a site for further monuments and memorials. In addition to the siting of political statues it became a focal point during royal coronations and jubilees. That it had become sacralized is indicated by comments made in response to the decision to reorganise the square in the late 1940s. Throughout the nineteenth-century Westminster had endured some of the densest traffic in the metropolis. By the mid-twentieth-century, with the demands of the motorcar and the decision to site the 1951 Festival of Britain on the South Bank, it was decided to alter the road layout to better enable the movement of traffic. When the future of the square and its various commemorative memorials was brought before parliament the Government was instructed to pay heed to the fact that this was 'the site of the very heart of the Empire' and needed to be treated with due deference (Parliamentary Debates, 1948-49: 470).

The architect George Grey Wornum (1888-1957) was responsible for the design whereby the central island was considerably enlarged to allow for the most extensive ' 'weaving' lengths for traffic on all four sides of the square' (Wornum, 1949: 137). Internal footpaths were created along the north and west angles and the statues were shifted from the centre to the periphery. The bulk of the central space was taken up by a grass clearing 'to provide a worthy pedestrian approach across it from the north side of the Square to the Abbey' (Wornum, 1949: 137).

This represents a modification to the initial phase of sight sacralization as conceived of in the nineteenth-century, thus indicating that it is not necessarily an exclusively linear process. When the Wornum scheme was implemented pedestrian access to the square was deliberately restricted on grounds of safety. Its remoteness may actually emphasise the sense of sacralization: rather than providing for its general use the design instead focused on 'the main needs of the public for viewing processions' (Wornum, 1949a). The grandiose temporary stadium erected in the square for the coronation of Elizabeth II in June 1953 testified to this (LMA 96.0 PAR: DS1322). However, to this day and under ordinary circumstances, the square remains underused and inaccessible.

Reversing the dominance of traffic and increasing pedestrian access is central to a far-ranging scheme initiated in 1996 under the title *World Squares for All*. Attesting to the marked shift in attitudes towards the urban environment, this project was in response to a study commissioned by Central Government and Westminster City Council. A multidisciplinary team of architects, urban and landscape designers and transport planners headed by Sir Norman Foster and Partners addressed the area linking Trafalgar Square, Whitehall and Parliament Square.[1]

Foster was of the opinion that:

> The London of the postcards is the nucleus of Britain, the most precious site in the land... Yet the innate harmony of Westminster is today invisible. Although pockets are well known and loved, the pieces do not fit together, severed by traffic arteries. There are more barriers than links. Sadly, the settings for some of the finest buildings are so appalling that they cannot be appreciated. To ignore the paucity

of space and allow traffic to rush past them is a national disgrace. (Foster, 1997: 5)

His reference to 'the London of the postcards' is one manifestation of the penultimate stage of sacralization: '*mechanical reproduction* of the sacred object' (MacCannell, 1976: 45). This had already begun in the nineteenth-century with the dissemination of souvenir photographs of the main sights in the capital by such firms as York & Son (NMR). This process of replication and the concomitant transformation of a sight into a mass-tourist attraction continues to the present day. The *London Eye*, a colossal Ferris wheel with a diameter of one hundred and thirty-five metres designed by Marks Barfield Architects, was erected in 2000 on the south bank of the River Thames opposite the Houses of Parliament (www.britishairways.com/londoneye). It provides an additional layering of marking, framing, elevation and means of reproduction both social and mechanical to this and other sacred sites in the centre of the metropolis.

In the above quotation, Norman Foster ennobles this central area as 'the nucleus of Britain, the most precious site in the land'. In spite of this, however, the process of sight sacralization has broken-down: the marking and framing of individual sites has led to fragmentation. To use the language of MacCannell, the number of barriers or 'physical divisions' has become overly excessive whilst the 'interstitial corridors' are so full of traffic as to invoke an expression of disgust: it 'is a national disgrace'. Foster's objective is to re-frame these sacred sites in a more holistic manner in order to make the 'innate harmony of Westminster' visible. This can be considered to represent the beginning of the third stage of sacralization: '*enshrinement*'.

Enshrinement

In MacCannell's opinion '*enshrinement*' occurs when the 'framing material' itself becomes integrated into the sight that has been marked off and elevated (MacCannell, 1976: 45). The area addressed by the *World Squares for All* project, as well as incorporating a World Heritage Site, consists of four Conservation Areas and over 170 listed structures, more than 30 of which are ranked Grade I (Masterplan, 1998: 21). It is, in its entirety, a 'sacred site' rather than a series of discrete clusters of 'buildings of outstanding or exceptional interest' (National Audit Office, 1992: 7).

This, a definition of a Grade I listed building, is reserved for structures 'of particularly great importance to the nation's built heritage' and 'likely to be of international significance' (Suddards and Hargreaves, 1996: 47-8). Included within this category are St. Margaret's Church, Westminster Abbey and the Palace of Westminster. Moreover, in 1987, these buildings achieved the status of World Heritage Site, for being illustrative, among other things, of 'significant stages in human history' (UNESCO 1972: 12-15; Number 426, 1987). This double inscription represents a further aspect of the *'naming phase'*. MacCannell stresses that, before this occurs, 'a great deal of work goes into the authentication of the candidate for sacralization... Reports are filed testifying to the object's aesthetic, historical, monetary, recreational and social values' (MacCannell, 1976: 44).

It has been observed that the 'boundaries' of many World Heritage Sites 'are inconsistent and are generally acknowledged as needing reviewing' (Suddards and Hargreaves, 1996: 70). Dr Christopher Young, Head of World Heritage and International Policy at English Heritage, states that the borders of the Westminster World Heritage Site were 'drawn very tightly' around the buildings and that, in retrospect, this has proven inconvenient. He believes 'it likely that the World Heritage Site Management Plan for Westminster, on which work is likely to commence shortly, will want to re-open the question of boundaries' (Young, 2001). As currently configured they divide it into two parts, with the Palace of Westminster in one section and the area around Westminster Abbey in another. It has been remarked that this arrangement 'has the curious result that Parliament Square with its statues of statesmen... [is] excluded from the site, despite being an integral part of the immediate setting of the Palace and the Abbey' (ICOMOS UK, 1995: 143).

The third and final phase of the £50 million *World Squares for All Masterplan* seeks to rectify this by removing traffic from the south side of Parliament Square in order to 'create an improved and appropriate setting for the World Heritage Site... [which] at present... is divided by heavy traffic and poor materials, with insufficient space for pedestrians' (Masterplan, 1998: 58-61 & 76). In the unlikely event of this being realised it would reverse the schemes of both E.M. Barry and Grey Wornum: the division between the Abbey and Parliament Square would be elided and enshrinement would occur.

The necessity of this process is demonstrated by the fact that Parliament Square is *already* linked with Westminster Abbey by being included within the same Conservation Area (reference number CA 20). It is in other words 'an area of special architectural or historic interest the character or appearance of which it is desirable to preserve or enhance' (Planning Act 1990: Section (69)(1)(a)). Such a designation is intended to address 'the quality of townscape in its broadest sense as well as the protection of individual buildings.' This includes recognition of 'the historic layout', 'particular 'mix' of uses', 'vistas along streets and between buildings; and on the extent to which traffic intrudes and limits pedestrian use of spaces between buildings' (PPG 15, 1994: 4.2). It is therefore clear that, for the historical associations of a monument to be sustained, the milieu of which it forms a part needs to be treated with sensitivity (Sanpaolesi, 1972). Or, to put in MacCannell's terms: 'Advanced framing occurs when the rest of the world is forced back from the object and the space in between is landscaped' (MacCannell, 1976: 45).

Profane Irruptions

It is, however, extremely difficult to force back the rest of the world in the construction of built heritage, as the example of Parliament Square has demonstrated. Furthermore, a prerequisite of a World Heritage Site is that it evinces 'an important interchange of human values' (UNESCO 1972: 12-15; Number 426, 1987). This means that it is very likely to still remain within an area of cultural, political and commercial exchange. It has been observed that 'sacred places exist in sacred landscapes, alongside, or nested within, secular places and secular landscapes' (Saunders, in Hubert, 1994: 172). Such locations are therefore often extremely difficult to manage. This is especially apparent with an urban environment like Westminster where a myriad of opposing factors converge upon a very diverse site that is subject to many conflicting demands.

The practice of sacralization represents an attempt to resolve this. As such it is an on-going process: *re-*sacralization is necessary in order to protect sacred sites from sacrilegious and worldly infiltration. There is a constant interposition between the contradictory requirement for imperative links and requisite barriers. The latter is essential in order to obviate the threat of desecration. In the case of Parliament Square this was made shockingly apparent during riots that occurred on May Day 2000. Every memorial, including the statue of the war-hero Winston Churchill (1874-1965) by Ivor Roberts-Jones (1913-96) of 1973 and Sir Edwin Lutyens's (1869-1944)

Cenotaph of 1920, was clambered upon and scrawled with obscenities by anti-capitalist rioters (Stillwell, 2000). At the time of writing, almost exactly one year hence, these monuments have had to be encased in wood in order to forestall a repeat of this profanity (Sutcliffe, 2001; Cummins, 2001).

The process of sacralization outlined in this chapter is crucial to the precept that the 'historic environment' is a central component of the tourism industry (English Heritage, 2000: 33). By being at an 'interchange of human values' it is axiomatic that attitudes towards sacred sites are subject to change. Successive generations will have interpreted and evaluated such domains of memory in a continual search for the 'original spirit of place' (Shackley, 1998: 194-5). However, just as the process of sacralization is never static, so too is the '*original* spirit of place' subject to reinterpretation. The 'authentic' aura of Westminster might conceivably be the '*cloaca* of narrow, tortuous, shabby, stifling, and malodorous streets' commented on so scornfully by G.A. Sala in the 1890s. That this is *not* the case confirms the fact that the built heritage needs to be sacralized in order for it to be both seen and protected. The outside world incessantly encroaches and the manner in which symbolic spaces are shaped– how 'the present frames up its history' (MacCannell, 1976: 88)– reflects the perceptions and priorities that each present places on its past.

Note

1. The Masterplan is managed by Westminster City Council and seven other study partners: the Government Office for London; the Department for Culture, Media and Sport; the Traffic Director for London; London Transport Buses; English Heritage; the Royal Parks Agency and the Parliamentary Works Directorate (DETR, 1999).

References

Anon (1827), 'Funeral of Mr. Canning' *The Times*, 17 August 1827. University of Nottingham Library, Department of Manuscripts and Special Collections: Os C 46.
Anon (1867), 'Parliament Square', *The Builder*, Vol. 25, 1287, 5 October.
Ashworth, G.J., Larkham, P.J. (1994), *Building a New Heritage: Tourism, Culture and Identity in the New Europe*, Routledge, London.
Baldwin, P., Eden, R. and Pook, S. (2000), 'Rioters dishonour war heroes', *Daily Telegraph*, 2 May: 1.
Blackwood, J. (1989), *London's Immortals: The Complete Outdoor Commemorative Statues* Savoy Press, London.
Brown, A. (1997), *'People Before Traffic' Vision for top London Landmarks*, Press Association Newsfile, 5 November.

Cannadine, D. (2000), 'The Palace of Westminster as the Palace of Varieties', in C. Riding and J. Riding (eds.) *The Houses of Parliament: history, art, architecture*. Merrell, London.
Croker, J.W. (1828), *Letter to Denison*, 18 February. University of Nottingham Library, Department of Manuscripts and Special Collections: Os C 48.
Crook, J.M. and Port, M.H. (1973), *The History of the Kings Works Volume VI 1782-1851*. HMSO, London.
Cummins (2001), 'The riot shields', *The Mirror*, 30 April: 17.
DETR (1998), *Prescott backs Masterplan to turn London's biggest traffic islands into places for people*, Department of the Environment, Transport And The Regions, Press Notice, 20 August.
DETR (1999), *Progress on Masterplan for Trafalgar Square welcomed*, Department of the Environment, Transport and the Regions, Press Notice, 10 March.
DETR (2000), *Better protection for Conservation Areas*. Department of Environment Transport and The Regions, Press Notice 135, 13 March.
Dresner, D. (ed.) (1995), *Collins Paperback Encyclopedia*, Harper Collins, Bath.
Edwards, J. (2000), 'This was their vilest hour', *The Mirror*, 2 May: 1.
English Heritage (2000), *Power of Place: the Future of the Historic Environment*, English Heritage, London.
Factsheet 48. C.C. Pond (1987) (revised C. Sear, 2000). *The Palace of Westminster*, House of Commons, London.
Foote, K. (1997), *Shadowed Ground: America's Landscapes of Violence and Tragedy*, University of Texas, Austin.
Foster, N. (1997), 'The New Golden City', *The Guardian (Features)*, 19 April: 5.
Hubert, J. (1994) 'Sacred beliefs and beliefs of sacredness', in D. Carmichael et al., *Sacred Sites, Sacred Places*. Routledge, London and New York.
ICOMOS UK (1995), *The English World Heritage Sites Monitoring Reports*, London.
LMA. London Metropolitan Archive.
MacCannell, D. (1976), *The Tourist: A New Theory of the Leisure Class*, Macmillan, London.
Masterplan (1998), *World Squares for All Masterplan*, Foster and Partners: London.
Meath, Lord (1901), 'The Enclosure in Broad Sanctuary', *The Times*, 17 July. Public Record Office, WORK 11/53.
Mullan, J. (2001), 'A brief history of mob rule', *The Guardian (Saturday Review)*, 28 April: 1-2.
National Audit Office (1992), *Protecting and Managing England's Heritage Property*, HMSO, London.
NMR York & Son archive held at the National Monument Record, Royal Commission for Historic Monuments of England (RCHME), London.
Orbach, J. (1987), *Victorian Architecture in Britain*, London: A & C Black.
Parliamentary Debates (1948-49). Parliament Square Bill- Second Reading, *Parliamentary Debates 1948-49*, House of Commons, Volume 470: 464-485.
Planning (Listed Buildings and Conservation Areas) Act 1990.
PPG 15 (1994), *Planning Policy Guidance: Planning and the Historic Environment*, Department of the Environment; Department of National Heritage.
Prior, W.H. (1866), 'Memorial Drinking Fountain, St. Margaret's, Westminster- Mr. S.S. Teulon, Architect', *The Builder*, 27 January: 65.

PRO WORK. *Records of the successive works departments, and of the Ancient Monuments Boards and Inspectorate, Office of Works and Successors: Statues and Memorials: Registered Files*, Public Record Office, London.

Sala, G.A. (1872), *Gaslight and Daylight, with some London scenes they shine upon*, Tinsley Brothers, London.

Sala, G.A. (1894), *London up to Date*, Adam and Charles Black, London.

Sanpaolesi (1972), *Preserving and restoring monuments and historic buildings*, UNESCO, Paris.

Schwarz, C. (ed.) (1997), *Chambers Giant Paperback Dictionary*, Chambers Harrap Publishers Ltd, Cambridge.

Shackley, M. (ed.) (1998), *Visitor Management: Case studies from World Heritage Sites*, Butterworth-Heinemann, Oxford.

Simpson, J.A. and Weiner, E.S.C. (1989), *The Oxford English Dictionary*, Oxford University Press, Oxford.

Stillwell, J. (2000), 'Anarchy thugs riot in Central London', Photograph. *The Times*, 2 May: 1.

Suddards, R.W., and Hargreaves, J.M. (1996), *Listed Buildings: The Law and Practice of Historic Buildings, Ancient Monuments, and Conservation Areas*, Sweet & Maxwell, London.

Sutcliffe (2001), 'Churchill goes under cover as London prepares for May Day riots', Photograph *Sunday Telegraph*, 29 April: 1.

UNESCO (1972), *Properties inscribed on the World Heritage List: Convention concerning the protection of the world cultural and natural heritage*, WH C.2000/3 Jan 2000.

Wheatley, G. (1997), *World Heritage Sites* (ed. Peter Stone), English Heritage, London.

Wilding, N. and Laundry, P. (1972), *An Encyclopaedia of Parliament* (revised 4th edition), Cassell, London.

Wornum, G. (1949), 'Replanning of Parliament Square', Grey Wornum, F.R.I.B.A., Architect, *The Builder*, Volume CLXXVII, Number 5554, 29 July: 137-139.

Wornum, G. (1949a). *Note of a meeting held in Room 420, Lambeth Bridge House at 10.45 a.m. on 25th November, 1949*. PRO WORK 22/170.

Young, 2001. E-mail sent to author on 12 March 2001.

17 Folkingestraat, Groningen: the Heritage of the Jewish Ghetto

G.J. ASHWORTH

Among the many possible incidences of heritage dissonance those relating to the management of the genocide of European Jews has a special importance. The extent, recentness and completeness of the identification, transportation, imprisonment and ultimate murder of around 6 million Europeans, in the heart of civilised Europe, presents future generations with perhaps the single most prominent illustration of the difficulties and sensitivities of heritage interpretation. Almost all Europeans can identify with the victims, the perpetrators or the bystanders and aspects of this heritage potentially exist in almost all parts of the continent (Tunbridge and Ashworth, 1996).

The scale and seriousness of the topic renders it impossible to present it here with any semblance of completeness. Only a brief review of the types and incidence of holocaust heritage can be given here as a context and space permits only a single relatively small-scale local case to be described. This must stand as representative of many more and often larger cases but it does demonstrate the nature of the more general issues involved.

If people create senses of place largely through their identification with place-bound heritage then the Holocaust presents three particular problems. First, the community that created the heritage no longer exists at the place where it was created which means that an exclave heritage now exists among a different local community. Secondly, the interpretation of the heritage of the victims is potentially dissonant for those identified with the perpetrators. Thirdly, much heritage is used in whole or in part as entertainment. The sensitivity of the heritage of the Holocaust causes particular problems when it is used within tourism as an element in a leisure activity. Certainly almost all heritage is multi-sold and quite different markets can be served at the same locations. However Holocaust tourism, which is rapidly growing, imposes a particular necessity for the careful management of quite different consumer groups.

The Jewish Holocaust as Heritage

The term genocide was coined by Lenkin in 1944, and has been defined as 'actions committed with intent to destroy in whole or in part a national, ethnic, racial or religious group' (United Nations, 1949). The attempt to eliminate the Jewish people, regardless of their actual religious or cultural affiliation, in Europe between 1933 and 1945 has become the most well publicised, documented and politically influential case of genocide, allowing the word *Holocaust* to be appropriated to describe it. Its heritage falls into two main categories namely the paraphernalia needed for the process of identification, collection, storage, extermination and disposal of around 6 million people and secondly, the previous living environments of the Jewish communities concerned. The first is largely represented by the concentration/extermination camps and the second by the residential ghettos throughout the major cities and less usually in rural areas.

The use of many of the surviving or reconstructed camps as heritage is well established throughout what was then German occupied Europe. A number have become major tourist attractions and the former inner urban ghettos are frequently marked and any surviving major structures (especially synagogues and cemeteries) preserved as monuments. A major management problem however that renders Holocaust heritage especially dissonant results from the very success of that genocide. Simply there is a spatial discrepancy between the incidence of the atrocity and its survivors or those who identify themselves as inheriting the role of victim. The community whose heritage is presented no longer inhabits the ghettos or even, in any number, the countries in which the memorialised camps are located. Equally dissonant is that Holocaust heritage is now located among, or even inhabited by, those who have inherited the roles of either perpetrator or spectator. Thus the 'whose heritage? /which heritage?' questions are especially contentious and almost any answers will result in policies dissonant to some.

The other face of Holocaust heritage is the primarily Central European inner city districts from which the Jews were taken. The ghetto was not, of course, invented as part of the extermination process, but had existed for many centuries in many European cities as part of an urban economic and social segregation resulting from varying degrees of law, custom and preference. It was however in some cases reconstituted as a 'holding tank' prison between 1941 and 1944 (Gruber, 1992). For example, at Theresienstadt, (Terezin, Czechia) the eighteenth century planned town was converted into a 'model' ghetto, in part for use in external propaganda (Gilbert, 1998). In a few cases, most notably Warsaw, the ghetto was physically obliterated after the 1943 uprising of its inmates and now exists only as a marked trail. More usually, and most obviously in the largest European ghettos, in Prague, Vilnius and Budapest, they were merely emptied of their Jewish inhabitants and reoccupied by others (Ashworth & Tunbridge, 1999).

This discrepancy between sites and people presents the most difficult aspect of any subsequent interpretation. Simply the survivors, their descendants and those who identify with them are no longer there as a result of the very success of the extermination policies. The largest communities in need of commemorative heritage of the ghetto are in Israel, the state founded in the wake of the genocide, and in the United States, the home of a numerous and influential Jewish minority. Heritage creation and marking in the former ghetto can therefore either address visitors identifying with a Jewish past but living elsewhere or local residents whose identification with the heritage may be weak, ambivalent or even hostile.

Thus the Jewish ghetto in the European city presents a peculiarly complex, sensitive and potentially contentious heritage issue. The preservation, marking and development for tourism of the old ghetto as a whole presents a series of social, cultural, political and purely practical difficulties. These can now be considered in more detail. The choice has been made not to examine a large and well-documented case, such as Krakow, Prague or Warsaw. These have been well studied (see Gilbert, 1998, for Poland in particular and Ashworth, 1998, for the heritage management of Krakow-Kazimierz, the largest of the Central European ghettos). In addition the smaller case is more typical and more widespread throughout Europe, and presents many mundane issues of town planning.

Folkingestraat, Groningen

The medium-sized city of Groningen in the northern Netherlands is typical in that it had until 1941, like most commercial cities of continental Europe, a sizeable Jewish population (some 3000 or around 8% of the city's total). Tolerated under licence since the late sixteenth century, they had been freed from special legal obligations in the course of the eighteenth century. Although there were thus no legal requirement for Jews to reside in a specific area or indeed practice certain professions, nevertheless many of the City's Jewish population, as well as many Jewish businesses and shops, were concentrated around the South-western sector of the historic city. In the 1900 population register the two streets, Folkingestraat and Nieuwestad were respectively 61% and 55% inhabited by Jewish households. There was no formal ghetto, non-Jewish households also inhabited the area and

Jews lived elsewhere in the City but especially in the southern and western districts within walking distance of the synagogues. Of the 404 heads of household registered in 1900, no less than 162 described themselves as 'merchants' and a further 70 as shopkeepers.

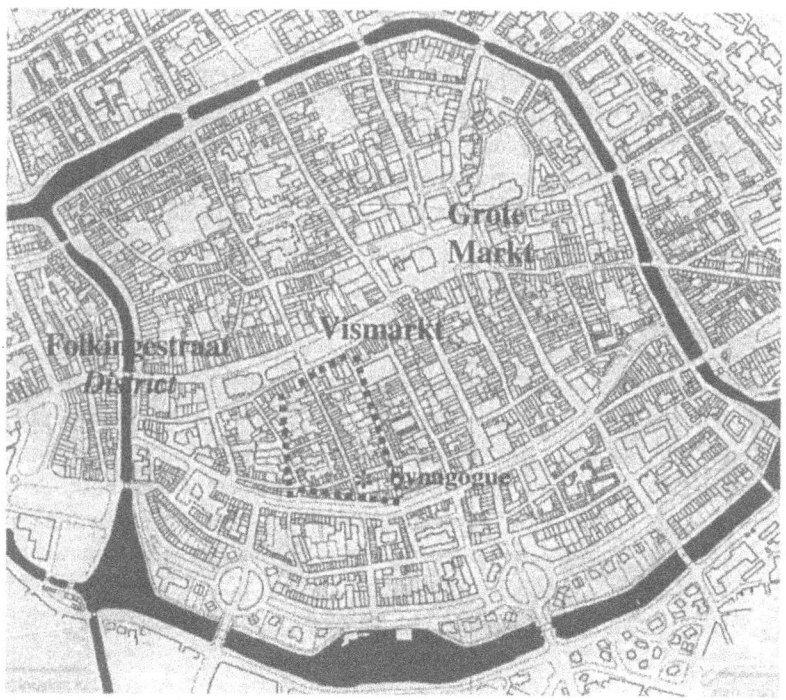

Figure 17.1 Location of Folkingestraat, Groningen
Source: G.J. Ashworth

The Folkingestraat area is part of the inner city near, but on the edge of, the main commercial centre. The community was large and prosperous enough to build an impressive synagogue at the southern end of the Folkingestraat at the beginning of the twentieth century. The neighbourhood supported a rich variety of Jewish religious, educational, sporting, cultural and social facilities and associations (Ast-Boiten and Zaagsma, 1996). This may suggest to outsiders a certain defensive 'ghetto mentality' and even discrimination by wider Dutch society. However within the Dutch 'pillar principle' of separate but equal patterns of life for each religious or political group, the official registration of religious belief and separate social and recreational provision was universal and not confined to Jews. It did however render the arrest and deportation of the community in 1941 both easy and comprehensive.

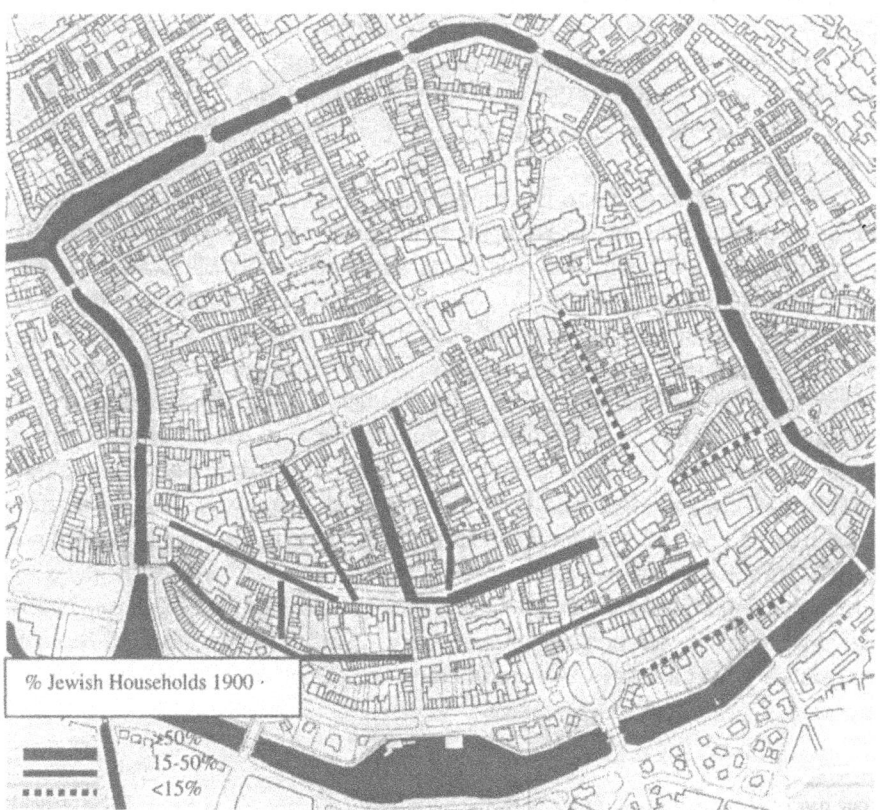

Figure 17.2 Jewish households in Groningen 1900
Source: G.J. Ashworth

The interpretation of the street and the surrounding district, presents two problems, one technical and one emotional. Both stem from the contemporary reality that there are no longer Jews here and thus the original consumers for whom the symbolic messages were written are not present to receive them and the present population either cannot or would rather not. Ironically enough the area of the former 'ghetto' is now partly inhabited by new ethnic minorities, notably from the Islamic Middle East and North Africa, thus the ghetto in the sense of an exotic ethnic ambience remained with new immigrants in a pattern of sequent occupance. The Folkingestraat developed as a relatively cheap rent district with shops and businesses serving both the local inhabitants (such as Arabic and Turkish butchers, delicatessens and video stores) and the wider city in search of a Middle Eastern 'souk' shopping experience. More recently there has been an urban renewal project (*Proefproject Stedelijk Herverkaveling*, 1994), the establishment of a shopkeepers organisation of 40 shops and particularly the establishment of new pedestrian circulation patterns consequent upon the building of the new Groningen Museum just to the south of the Folkingestraat. This has led to a rapid 'bohemianisation' as art galleries and boutiques have been added to the commercial mix.

The Jewish heritage is memorialised in the conserved synagogue, a national monument, some regular exhibitions on the vanished Jewish community and a standardised plaque marking. An important question is 'for whom is this heritage produced?'. The present Jewish population of the City, and indeed the country, is minuscule. Many local residents are likely to find the only remaining intelligible message either irrelevant to them, or even provocative, given the impacts of later Zionism on the Arab world. The relevance to the wider Dutch community is the uncomfortable memorialisation of collaboration, or just inaction, during the deportation. Finally the tourist market is unlikely to make much use of a ghetto product for a number of reasons. Foreign tourism to the City is modest in scale, and largely attracted by the seventeenth century architectural-heritage package to which the Folkingestraat has little contribution to make. Visitors specifically seeking out the Jewish heritage are rare and by far the largest group of foreign tourists are from Germany, who would thus be presented with the uncongenial message of complicity in genocide.

Whose Folkingestraat: Whose Heritage?

Much of the unresolved tension and resulting paradoxes inherent in this case stem from a common duality. In one sense the ghetto is a unique phenomenon because of the unique dimensions of the crime of the Holocaust. In another sense, however, it is archetypal of the cultural and ethnic mosaic reflected in a separate and distinctive built environment that could have housed any of the many possible communities. If the atrocity element were the only consideration then it would be relatively easy to accord paramountcy to the national and international memorial function. It was however such a widespread phenomenon throughout European cities even containing a majority of the population in some Polish cases (Gruber, 1992) that it merges into more mundane issues of the local revitalisation and renovation problems of inner city districts. It is this clash of the sublime and the mundane, the sacred and the secular, the international and the local that provides much of the complexity now facing the city planners as they embark upon renewal in such districts.

The relationship of the heritage to tourism is similarly complex. As all heritage is a customer created and defined multi-sold product, it has been brought into being to serve various purposes and is sustained by various users. Tourism is thus one of many uses and cannot be considered except in this context of multi-use. Equally heritage produced for local markets for purposes of local place identity will be quite different from that consumed by tourists. All heritage tourism is thus to an extent marketing an exclave heritage. A different product is sold to each of the different markets. This may or may not matter but certainly requires careful management of resources, products or markets.

Three further complications have also been added in this case. First that the heritage is that of atrocity raising the requirement for sensitivity to passionately felt and enduring emotions to a higher level. Secondly, the heritage forming the basis of the heritage resource is specifically that of genocide, which by definition removes the victims, the most important group likely to wish to identify with it and conveys an unwelcome identification upon another group, the perpetrators. Thirdly, there is here as in many such areas a cultural and thus heritage succession. One cultural group has been replaced by another and the heritage, like the ethnic succession, has become multilayered. The interpretation of any one layer is not only a partial reflection of the complex heritage and identity of this place; it is likely to impact upon the others.

There are thus many arguments for a policy of inaction or even deliberate induced collective amnesia. The Netherlands, in common with most German occupied Europe developed a tacit policy of amnesia about the events of 1940-1945 in order to preserve social cohesion for the immediate post-war generation. Some individuals and institutions had actively collaborated; some actively resisted and most had remained passive. The preservation of political harmony and avoidance of social discord encouraged an orientation towards the future not a dwelling upon the past. However such pasts cannot just be erased from human memory for more than very short periods: the inheritance cannot in practice be rejected. Again, as in much of occupied and neutral Europe, interest in this period revived 50 years later among a new generation. Identification with the victims, the perpetrators or the spectators will exercise an enormous fascination and curiosity that will generate a demand from both locals and tourists for motives that are a combination of education and entertainment and which cannot be separated into acceptable and unacceptable categories. The choice is not between the two clear alternatives of either interpreting such heritage or ignoring it. The choices lie in how it is so interpreted and for what purpose but even here managers have less influence on how such interpretations are received than they might wish.

Groningen-Folkingestraat is only one of many memorialised ghettos to the Jewish Holocaust in Europe which in turn is not the only genocide to haunt human pasts. This is neither an exclusively Jewish nor Dutch problem nor is it confined to Europe. If all heritage is potentially dissonant in some way then all heritage management must be aware of its potential in this respect. In these senses the tensions in the local case considered here apply to any relationship of users to heritage, it is only that the subject matter renders them more visible, significant and sensitive.

References

Ashworth, G.J. (1998), 'Jewish culture and holocaust tourism' in M. Robinson (ed) *Culture and Tourism*, Centre for Tourism Studies, University of Northumberland, Newcastle.

Ashworth, G.J. and Tunbridge, J.E. (1999), 'Old cities, new pasts: heritage in the transformation of Central European cities', *Geojournal* Z.Kovacs (ed) pp.105-16.

Ast-Boiten, L. and Zaagsma, G. (1996), *De Folkingestraat: geschiedenis van de joods gemeenschap in Groningen* Forsten, Groningen.

Gilbert, M. (1998), *Holocaust Journey: travelling in search of the past*, Penguin, Harmondsworth.

Gruber, R.E. (1992), *Jewish Heritage Travel: a guide to Central and Eastern Europe*, Wiley, New York.

Lenkin, R. (1944), *Axis Rule in Ooccupied Europe*, Howard Fertig, New York.

Tunbridge, J.E. and Ashworth, G.J. (1996), *Dissonant Heritage: the management of the past as a resource in conflict*, Wiley, London.

United Nations (1949), *Yearbook of the United Nations 1948-9*. Columbia University Press, New York.

CONCLUSION

18 The Experience of Heritage Conservation: Outcomes and Futures

G.J. ASHWORTH

The preceding text demonstrates clearly that despite continuing disagreement concerning what to conserve, why, how and for whom, there is considerable investment of attention, energy and finance in the process of heritage conservation and interpretation. The question 'whose heritage?' is one of the oldest questions raised in discussions about the selection, preservation and interpretation of the conserved built environment. Answers to it were implicit in the establishment of the organisational and legislative structures for the preservation of buildings and areas from their inception. However, it began to be posed explicitly, usually by academic observers, once the results of the practice of built environment conservation became evident on a substantial scale in cities (see for example, Tunbridge, 1984). Answers to the question have dominated heritage discussions since, have changed with mutations in the preoccupations of society and intellectual fashion and have nearly always received multiple answers.

An objective of this conclusion is to explore the themes that emerge from the studies and question whether they do indeed represent a common definition of heritage values. The point was made in the introduction that the heritage of the built environment is not the result of haphazard survival, but relates to intentional choices to create, maintain and preserve selected 'senses of place'. It follows that although heritage values are not locally defined, they may be more readily recognised at local level. There are wider concerns that are generally held that underpin the concept and unite different presentations of heritage in different locations. The studies demonstrate that there is a correspondence in these concerns that transcends the national. In drawing together points of comparison from these studies and emphasising some differences, it is hoped this book helps to shed some light on the international currency of both the term and the practice of heritage conservation.

The National Experiences

The three national sketches of the evolution of a concern for the relict built environment provided similar answers. In all three national heritage programmes were adopted by national governments in support of perceived national needs at the time. The simple link between the political philosophy of nationalism and the creation of the entity, nation, is self-evident but the past has been utilised in different ways, many of which are by no means self-evident, as the most used instrument, for forging, delimiting and strengthening this link. Ideas of legitimation (Habermas, 1973), cultural capital (Bourdieu, 1990) and dominant ideology (Abercrombie, et al., 1980) have all been used to explain the strength and significance of this link but not its subtlety.

The Swedish claim to an historical primacy in preservation legislation was quite directly related to the historical circumstances in which Sweden found itself in the seventeenth and eighteenth centuries. In particular in relation to its ongoing territorial conflict with Denmark and its political and military ambitions on the Southern and Eastern shores of the Baltic created the need for identification of the Swedish crown with a distinctive population group that could identify with it. There are elements in this identification, that re-occur even more strongly elsewhere, of not only a self-conscious national solidarity but also an historical destiny which legitimates territorial claims. The idea of the 'Goths' making a distinctive contribution to European civilisation which was then diffused by them through the Nordic and Baltic lands was a form of 'manifest destiny' that predated by two centuries the coinage of the term. In our argument the important question, 'manifest where?' is answered, 'in our history'. If this seventeenth century idea now seems jejune, not least to contemporary Swedes, then it is worth recalling that visitors to Athens airport were, until very recently, greeted by an official banner declaiming, 'Macedonia is Greek: read history'.

Although both The Netherlands and Britain were from the seventeenth century what we would now term, 'global players', neither can reasonably claim to have developed a heritage justification, supported by formal legal structures, for their global commercial and political expansion at the time. In both the idea of the 'golden age' (in Britain normally located in the sixteenth, and in The Netherlands seventeenth, centuries) was a creation of the late nineteenth century well after the global expansion had occurred and when it was, if not in retreat, at least under perceptible threat.

The timing of official concern for cultural heritage can be related to both specific needs and threats but also to more vaguely felt and experienced trends. The coincidence of the timing of legislation in countries across Europe seems to suggest the operation of continent or even to an extent world-wide trends. The countries that experienced first and most strongly such trends were thus most under threat and consequently were those that pioneered conservation. These are represented by our national cases: Britain, the first urban-industrial country of the modern age, The Netherlands, with its tradition of global commercial involvement since the seventeenth century: and Sweden, whose early industrial development was a precursor of the eighteenth century industrial revolution.

The second half of the nineteenth century witnessed the consequences of rapid industrialisation, urbanisation and social change. A corollary of progress to the new is the abandonment of the old and this can result in nostalgia, etymologically the pain experienced through loss originally of home. In this model, heritage is invoked as a reaction to a perceived threat not so much as an attempt to return to inhabit a re-creation of a vanishing or vanished past but to satisfy contemporary needs for adapting to, and becoming reconciled with, change. The cry, 'Save our heritage', with 'our' being defined in national terms, implies that the two questions, 'save from what?' and 'for what purpose?' have already been answered and are thus self-evident.

The objective arose to preserve some relics of what was seen to be a vanishing or already vanished way of life as a reminder and a legacy. The threat is modernity and the reaction is not so much resistance to change in a retreat to the past, as the rendering acceptable of such change through control of its pace, channelling of its directions or shaping some compensatory balance.

Curiously in both The Netherlands and Britain, the 'gothic', originally a derisory term of classicists for the cultural barbarism of the 'dark ages', was rediscovered as the ages of faith, order and certainty at the moment when these characteristics were seen as being weakened by the new economic realities and their social consequences. Ruskin and de Stuers were the castigators of the short-sighted greed of modernity and Pugin and Cuypers, the re-creators of the medieval world as counterpoise to it.

Again Sweden provides one of the most pervasive of these defensive strategies in the place-name eponymous *Skansen* movement. This gave rise from the last quarter of the nineteenth century, to folk museums where not only vernacular buildings and artefacts but also craft activities and ways of life were preserved, reconstructed and represented. The 'vernacular', the 'everyday' and the 'folk', and, underlying these, the sets of moral values that these were assumed to represent, were all seen as being under threat and worthy of preservation and transmission to the present and the future. However it is worth noting that museums remained specially designated sites separate from the 'real' world outside, with which they interacted only as a temporarily experienced contrast not as an instrument of reform of that world.

More recently, and a century later, apprehension about the consequences of a new wave of economic, political and social globalisation has resulted in a similar reaction manifested in very similar ways. Economic and political globalisation, translated in some European countries as 'Americanisation', has reinvigorated a deliberate search for a localism that can be intentionally fostered as a counterpoise to global social homogenisation. The Skansen movement of the late nineteenth century is echoed in the popular idea of the 'ecomuseum' of the late twentieth century : the moral imperative of the craft and the vernacular now become the equally morally desirable harmony of economic and cultural activities in the environment at the scale of the locality. Such ideas are also implicit in the emergence of the concept of the 'conservation area', not as a museum but to be lived and worked in. This is made quite explicit in a major new inter-ministry initiative for landscape and city preservation in The Netherlands (the 'Belvedere Report', 1999). Here it is quite clearly stated that the programme's primary goal is the need to enhance feelings of local identity though encouraging the preservation of the cultural history of landscape and cityscape. This is not a rejection of economic globalisation, in which the Dutch companies and financial institutions are major operators and Dutch society a major beneficiary, so much as a search for some compensatory balance elsewhere.

If the answer to the 'whose heritage' question is 'our nation's' then a practical aspect of this is the ownership of those cultural properties that symbolise and transmit this collective heritage. Buildings, artefacts and even sites are owned by somebody, which raises the question, 'whose heritage property is it and what rights does this ownership entail?' In all three national accounts there is a long struggle to resolve the question of the ownership of cultural property. Liberalism had developed as an assertion of the individual's rights to property, its untrammelled enjoyment and even its ultimate disposal, against other especially feudal or despotic claims. The idea of heritage implies recognition of some collective rights and thus responsibilities which take legal priority over individual rights. This historical struggle of private against public ownership, and individual against collective rights lasted for almost a century in each of the cases discussed. Although this conflict is far from being resolved in North America, a broad consensus was achieved in Europe which struck a balance by accepting a multiple ownership, restricting some individual rights in compensation for some collective subsidy. In all three countries this consensus was a compromise enshrined in law, planning practice and informed public opinion.

Heritage, Identity and Urban Regeneration

The three cases grouped under the heading heritage, identity and urban regeneration all examine the way conserved historicity has contributed to much broader planning objectives at the scale of the city district. The conservation of the heritage of the built environment is viewed as a planning process in which new valuations of historic forms are part of a process of urban change.

The critical shift in focus is from the preservation of physical entities, whether individual buildings or morphological patterns, for their intrinsic worth, to the conserved historicity of areas and ensembles, to which have been ascribed extrinsic qualities. Such historicity may be expressed in different ways but is usually only one element in a much wider set of area based variables. Central to each of the city district examples is the question, 'who capitalises on the locally released benefits of heritage?' and its corollary, 'who pays for this?'. These questions are manifest most starkly, and are most well documented, in the dilemma inherent in the process of gentrification based upon heritage values. Old, low-rent urban neighbourhoods in need of physical renovation, and perhaps also economic and social revitalisation, offer an opportunity for a return upon investment to be gained through realising the heritage values. Such investment is usually too high to be borne by public authorities alone but also the risk is too great for private investment to occur without public involvement, stimulation or quarantee. The costs of this process may be born in part by the original residents who are displaced by rising rent levels. The dilemma is that what may be condemned as undesirable speculation or the utilisation of public subsidies for enhancing private profits, is a necessary means of harnessing private investments in a restoration task whose dimensions would be well beyond the means of public authorities and whose benefits accrue to both individual investors and the collective interest. Thus local authorities not only need gentrification: they must seek out possibilities for it to occur. Neither Göteborg's Haga district, which is principally residential, nor Nottingham's Lace Market, which is a mix of residential and commercial gentrification, could have been restored without the possibilities for private profit.

The case of Groningen's Waagstraat involved an even more direct and deliberate replacement of one set of buildings and their functions by another. It was not, as in the other cases, that old, newly valued, forms received new functions but that the newly shaped historicity required new forms as well as new functions. International style architecture of the immediate post-war period was replaced by post-modern italianate design building and office and local government functions were largely replaced by retailing, catering and residential uses. The heritage question posed and answered, at least at that time was 'what heritage image does the city wish to project through its city centre?' It is notable in itself that the main function of the city centre has become, or maybe reverted after an aberrant post-war period, iconic. The discussion, within and beyond local government circles was not, 'what are or should be, the functions of the city's heart' but which symbolisms should be selected to portray which self-consciously shaped identity. Who chooses which heritage in support of which identity to represent which and whose city? It is worth noting that these processes of reassessment or restructuring of the past, is continuing in Groningen as recently (2001) the City Council's plans for another area of the city centre were soundly rejected by popular referendum, fought very largely on the issue, 'whose city is it?'

The Heritage Site as Attraction

The three cases grouped under the theme heritage sites as attraction shifts the focus not only of scale from the multifunctional city to the heritage site but also introduces a major consumer of the past, the tourist, and economic justification for its conservation, tourism. The introduction of tourism has two consequences first it is the archetypal economic heritage product generator and consumer and thus raises a specifically economic dimension to the discussion of heritage values. Secondly, the tourist represents in a visible form a non-local valuation of, and claim upon, the site and thus consequent demand to use it. Answers to the 'whose heritage?' question may now be 'the locals' heritage' or the 'visitors' heritage'. Conflict may occur either if these are different heritages or if they are the same heritage. The first type of potential conflict stems from different interpretation packages being sold to different markets on the same site: the second from the pressures of joint demand upon scarce resources including space.

Such potential conflicts are generally resolved by prioritisation. The question, 'whose heritage takes precedence?' can be answered in favour of either tourists or local communities. Although it is superficially attractive, and more acceptable to local political systems, to accord primacy to local interests in cases of conflicting demands for heritage, this creates both philosophical and practical difficulties. Automatically assuming that local power elites have a stronger claim than national ones, is, if followed to its logical extreme, the 'road to Bamyan'. It is also mundanely almost impossible to implement through discriminatory management policies.

It is clear in the two studies of the large historic houses, Gunnebo and Wollaton Hall, that tourism as an economic activity is only one, and probably not the central, justification for the existence of the sites. Visitors are welcome for their economic contribution to maintenance but equally as much for the justification their presence contributes to the educational and cultural programmes and to the political decision to maintain and operate the establishments. Nevertheless the need to accommodate visitors, and the desire to maximise their financial contribution results in pressures that threaten the perceived authenticity of preservationist decisions. The solutions adopted in these cases have produced completely different results, one prompting an extensive new build, albeit utilising authentic craft designs and methods, while the other will attempt to absorb modern features such as a lift within the original structure, but in both cases the presentation to the visitors can be seen by some as compromised.

The small fortress towns, exemplified by two Dutch cases, which could have been replicated throughout Europe, have the complication of being not merely assemblages of designated historic monuments but also towns, however truncated and subordinated these other urban functions have been. There is no real doubt that the 'whose heritage?' question is being answered, 'the nation's' or even 'the world's' as represented by the tourist but the answer 'us citizens' is at least a complication.

Heritage as a Strategic Policy Option

Heritage can be seen as a policy option for strategic development at various spatial scales. In practice aspects of historicity, including both fostered historic site associations and restored historic structures can be incorporated as key catalytic elements in development plans. The most important point is precisely that heritage is an option and as such can be rejected as well as accepted. Unlike the individual heritage attractions discussed earlier place bound heritage is neither an inevitable priority of place managers nor is the enhancement of historicity usually the ultimate goal of such strategies. It is only a means of achieving a much wider economic or social goal. As such the key questions to be addressed are what are these goals, what are the alternative strategies available, and more particular to our topic, what are the various roles heritage might play in any such strategy. The approach is therefore essentially comparative both in the sense of evaluation between alternative approaches and in the assessment of the competitive arenas that these places are entering.

The three cases considered illustrate the possibilities of failure as much as the preconditions for success. If heritage is treated as a local economic resource then, in so far as all places have a past, it is a ubiquitous resource. History may be place specific and unique but heritage is almost infinitely replicable. Indeed, the heritage industry can be regarded as a low skill, low capital investment and generally low entry cost option that is available to everywhere. It is thus surprising and deserving of explanation, that so many places enter into competition in such satiated markets so easy of entry where success is likely to be elusive. All too often heritage is invoked as a development option as a last resort when other options have failed or are clearly impractical. Both Bolsover and Nieuweschans have reluctantly turned to heritage as a result of previous economic failure and resulting social distress. In these circumstances the irony is that the need for success may be greatest in precisely those places where the chance of such success is probably lowest as evidenced by previous failure and the lack of other alternatives. It also cannot be assumed that the playing of the heritage card is automatically beneficial to all economic development. A heritage approach is as likely to repel as to attract new economic activities. Where there is a choice of pasts to select, the focus on more distant stories may only act to perpetuate recent grievances, and drive a further wedge between the local community and visitors. Heritage consciousness can be a synonym for a backward looking resistance to change that is not an attractive locational factor for economic enterprises that are not 'selling the past' nor residents who do not relish 'living in a museum'.

In Forsvik this perhaps matters less as the use of the previous economic activities, structures and associations to continue work as well as provide a heritage resource in support of a range of new tourist and educational activities is not only self-evidently the only option, it is probably provides sufficient economic benefit to support existing society. However the larger settlements of Bolsover and Nieuweschans could not rely upon heritage alone, even if it was as successful as the programmes hope. In all three cases a major role of heritage is the restoration of some degree of local self-confidence, in the larger towns as a necessary precondition for revitalisation in quite other fields rather than the generation of new heritage based export products.

Heritage and the Restructuring of Symbolic Places

Places in contrast to geometric spaces have values ascribed to them. Messages are encoded upon them and transmitted to those who can decode them. Heritage places in particular are potent conveyors of such messages but equally, given the passage of time between encoding and transmission, have particular difficulties of interpretation. As Raphael Samuels observed 'aesthetically as well as historically heritage is a hybrid, reflecting, or taking part in, style wars, and registering changes in public taste... in any given period conservation, and with it 'heritage' will reflect the ruling aesthetic of the day' (Samuels, 1994: 211). This quite simple idea creates heritage dissonances which in most instances are not especially significant. However the cases raised here are selected from the minority of instances where the messages emanating from heritage for one reason or another are dissonant to those receiving them. On a spectrum of seriousness, they may just be unintelligible or no longer relevant, or cause disquiet, offence and conflict. It is not only the content of messages that is important but also what they do not contain. All heritage by definition disinherits someone, in some way but in most cases this is limited in its effect or even trivial. However in other cases such disinheritance may ignore, marginalize or even deliberately exclude or stigmatise particular groups.

The 'whose heritage?' question is rephrased and extended in a new interrogative series. Whose messages are to be encoded and transmitted, in pursuit of whose objectives and who is the intended recipient? To each of these questions can be added the negative corollary, 'and whose not?' The analysis resulting from such an inquisition is only a preliminary for proceeding to the policy questions of 'what could, should or must be done?' Heritage can be reselected, new heritages being transmitted, de-selected through demolition, concealment or marginalization, re-interpreted to include alternative or additional narratives, managed for multiple consumer markets. In other words there are numerous heritage management options, all of which require a sensitivity to the existence of actual or potential dissonance and to the consequences of the projection, transmission and reception of messages.

Almost all the cases described under each of the thematic headings could on close investigation reveal some aspects of these dissonant elements. The Dutch fortress towns of the 'golden century' celebrate the triumph of urban, mercantile, bourgeois, protestant, capitalist, nationalism over imperial, Hapsburg, catholic, feudal, internationalism. The passage of time has rendered most of the distinctions largely irrelevant but are Catholics, socialists, pacifists to say nothing of Moslems, Buddhists and Hindus excluded or, at least, less likely to identify with it? The heritage of the industrial revolution, as in Nottingham's Lace Market or at Bolsover may well be viewed by those who worked and lived in it as a heritage of hard times best obliterated and forgotten rather than commemorated and memorialised. The fortress of Nieuweschans can be resurrected as a pleasant high amenity residential and tourist-commercial urban environment. However fundamentally it was created by the needs and technologies of an unpleasant activity, warfare, and specifically directed against the Dutch folk enemy, the Germans, who are now to be entertained in it as tourists.

However four cases were included specifically to exemplify aspects of this problem with particular contemporary significance. The discussion of Parliament Square London raises directly the question, whose heritage and what heritage should be reflected in this central symbol of the British state. It is a grander and better documented illustration of the same political symbolic role of heritage discussed in the Groningen Waagstraat. It is notable that the design of Parliament Square has been a continuing discussion that commenced around the 1840s with the construction of the major parliament buildings, but was continued with the rebuilding after the Second World War damage up to the present office extension building and new statuary. Continuing change only reflects changing answers to the question of whose heritage is representative of the changing idea of the British state. The case of the Vänersborg museum illustrates similar questions at a more modest scale, where issues of location and layout reflect contemporary views of exclusion and exclusivity. That these issues are readdressed for the modern audience emphasises the point that solutions are contingent not fixed, periodically renegotiated, not determined by narrative sequence.

The memorialization of unsavoury aspects of the past confronts heritage management with one of its most widespread, serious and intractable problems. Slavery has been selected from the long sad saga of history, to represent this type of heritage. Cities that profited from the Atlantic slave trade in particular, of which Liverpool is one example, are prevented from ignoring or marginalizing this experience because a substantial minority of its inhabitants identify in some way or another with the victims. They demand that what they regard as their heritage be included in the narrative alongside the hitherto dominant interpretations and memorializations of the City's heritage. There are three possible outcomes. At worst there is the presentation of conflicting, contradictory and divisive messages. Secondly there is simultaneous transmission of separate heritages to separate groups (what might be termed the 'Derry Museum solution', where two different marked tours are offered to 'orange' and 'green' visitors respectively). Finally at best, a composite heritage could be devised that somehow represents and satisfies all recipients as their heritage. In any event sensitive management is essential.

The final category of dissonant heritage to be illustrated is that of enclave heritage. Here the heritage buildings and sites relate not to the present population but to a society that is no longer present in the area. The heritage of the Holocaust of European Jewry, especially the urban ghettos is almost ubiquitous throughout continental European cities while the nature of the atrocity killed or displaced those who most strongly identify with it. This is thus the heritage of someone else not us. In the case of Folkingestraat, Groningen the local population may view the heritage as little more than an interesting exotic and essentially foreign episode in urban history or even as an unpleasant reminder of complicity, however passive, in genocide. The coincidence that much of the replacement population for the Jews was Moslem Arab or Turkish adds an extra dimension to the rejection of the heritage as not ours.

The Search for Answers

It is clear from the cases introduced and discussed that the 'whose heritage?' question recurs in similar contexts throughout the countries studied, but that there is no definitive answer. Or rather the answer depends upon who poses the question, in which context and with which objective. Further it is clear that the question will be continuously re-posed and re-answered in each successive time period according to society's changing priorities. Each generation must not only re-write its history; it must also recreate its heritage. The implications of this are that heritage, its selection and interpretation, will be in a process of continuous change. A major difficulty in implementing this simple idea lies in the resistance to change that is intrinsic in the processes of heritage creation and the nature of most heritage objects and sites. Such inertia is particularly evident in the heritage of the built environment which is by its nature physically robust. Inventories and especially the criteria used to create them, change slowly: the expert consensus about what is, and is not, worthy of recognition and perhaps even more so public opinion, changes even more slowly. 'New' sites and structures can be relatively easily designated to reflect new societies: local, national and world heritage inventories are growing rapidly and expanding their coverage over previously overlooked activities, groups, areas and periods. Change however implies more than mere addition. Very few structures or sites are being de-designated, previously valued artefacts and buildings destroyed, or even few now irrelevant interpretations are being consigned to the dustbin of history. The present has in effect been colonised by the values of the past as reflected in their heritage in the same way as the present is busy colonising the future with its values as projected by our current heritage. The world could be seen as being in danger of being littered with the relict heritages of past generations much of which now mean little to contemporary societies.

Finally and most tentatively the 'whose heritage?' question can be posed in a predictive way. If answers depend upon the needs of society then speculation about such contemporary needs should guide current heritage policies. An attempt to sketch the nature of contemporary western society and thus its requirements would lead the discussion into broad, uncharted and potentially stormy waters. However there would be little disagreement that western society is becoming more individualistic, atomised, pluriform, and multicultural however these terms are defined. This does not mean that heritage could or should be anything and everything: it does mean that the issues of inclusion and exclusion, of social division and cohesion and of individual identification and marginalisation, raised in the cases above, are likely to be critical elements in the policies of today for shaping the heritages of tomorrow.

References

Abercrombie, N., Hill, S. and Turner, B.S. (1980), *The Dominan Ideology Thesis* Allen & Unwin, London.
Bourdieu, P. (1990), *Distinction.* Cambridge University Press, Cambridge.
Habermas, J. (1973), *Legitimationsprobleme im spätkapitalismus*, Suhrkamp, Frankfurt am Main.
OCW, LNV, VROM, VW (1999), *Belvedere: Beleidsnota over de relatie cultuurhistorie en ruimtelijke inrichting. Bijlage: gebieden*, VNG-uitgeverij, Den Haag.
Samuels, R. (1994), *Theatres of Memory* Verso, London.
Tunbridge, J.E. (1984), 'Whose heritage to conserve? Cross cultural reflections upon political dominance and urban heritage conservation', *Canadian Geographer* vol. 28, pp. 171-80.

Subject Index

amnesia 204, 243
Anthonisz (Adriaen) 134
antiquities 14
archaeology 15, 18, 20, 77, 112, 114-15
    see also industrial archaeology
architecture 23, 42, 84, 92, 119, 209, 212-16 *see also* urban design
art 171 *see also* culture
    Nouveau 48
Aubrey (John) 14
authenticity 96, 101, 104-5, 110, 113-14, 116, 141, 146-47, 253

Bauer (Walter) 105
Bureus (Johannes) 30
Burra Charter 190

CADW 26
car parking 136
Carlberg (Carl) 104-5, 109-13
castles 18, 73, 75, 121, 151, 157, 158-9
    see also fortress
Charter of Venice (1964) 32, 104, 110
churches 15, 18, 32, 39, 51, 219, 224
civic
    consciousness 145
    trust 16-17, 21, 76, 80, 119
classicism 209, 249 *see also*
    Palladian
Coehoorn (Menno van) 134, 138
    Stichting 134, 135
commodification 26, 116, 142, 203
community 9, 60
conservation
    areas 19, 20-22, 25-6, 47-8, 73, 77-8, 80, 83-4, 91, 125, 150, 159-60, 168, 231-3, 250
    consciousness 3, 49

enforcement 21
legislation 14, 15, 18-24, 25, 32-3, 36, 41, 46-7, 77, 168, 247
terminology 190
Council of Europa 104, 189 *see also*
    Europa Nostra
country house 103-16, 117-30
cultural
    capital 248
    facilities 91, 250
    heritage 118, 181-2, 186, 190, 204, 243, 248
    policy 91
    programmes 105, 182-3
    quarter 202
    tourism 5, 107
Cuypers (P.J.H.) 46, 249

deindustrialisation 5, 83, 90-1, 152, 175
Disneyfication 145, 170
docklands *see* waterfronts
DoCoMoMo 16

Eberhard (Karl) 215-16, 217
economic
    change 13, 47, 166-7, 176-9, 193-4, 250
    regeneration 26, 153-61, 163-4, 173, 253-4
    viability 23
    English Heritage 17, 20-21, 79, 123-4, 126, 154, 157, 158, 234
entertainment 83, 136, 237
environment 79
    impact 38
    improvements 81, 125
    Europa Nostra 80 *see also* Council of Europe
European Walled Towns

Friendship Association 131
flâneur 210-11, 220
fortress 102, 131-44, 146, 151, 164-6, 178, 253, 255
Foster (Norman) 230-31

garden 40, 107-8, 114-15, 117, 121-2, 226
   History Society 21
gem cities 131
   Georgian 77, 120, 191-2 *see also* Palladian
   Regency Society 16
gentrification 25, 27, 52, 60, 257
ghetto 237, 238, 239, 240, 242
globalisation 6
Goths 29-30
   Gothic revival 15, 33, 249
Gustaf Adolph (King) 29, 30

Hazelius (Arthur) 35
Heemschut 46
heritage 32, 91, 101, 133, 145, 168, 190
   class 67-9
   conflict 8, 142-3, 238-9, 252-3, 256-7
   costs 15, 51, 116, 173
   dissonance 133, 141, 202-5, 243-4, 255-6
   see also multicultural, race, slavery, Holocaust
   employment 109, 111, 147, 161-2
   financing 8, 50-1, 65, 66, 80, 108, 123, 158-9
   industrial 26, 42, 58, 73-85, 123, 145, 175-87
   industry 3, 26, 146-7
   interpretation 6, 8, 88, 123, 125, 133, 142, 158, 189, 198-202, 207-21, 237, 241, 243-4 *see also* marking
   management 32, 103, 128, 190, 207, 252
   marketing 146, 223, 237, 255
   ministries 26, 135
   multicultural 5, 258
   planning 53, 88, 140, 145-7, 163-4, 173, 183-5
   promotion 103
   societies 16, 17, 33, 35, 36, 46, 77, 81, 181 *see also* national trust; civic Trust, CADW
   threats 24
   tourism 5, 13, 26, 51, 81, 141-2, 156-7, 170, 243
   trail 159, 170, 199, 201-2
Holocaust 189, 237-9, 256 *see also* Jews
housing 43, 51, 61-2, 78, 82, 118-21, 135-6, 171, 255
   Association 80

ICOMOS 232 *see also* UNESCO
identity 6, 57-8, 142, 189, 248, 251-2
immigration 6
industrial archaeology 146 *see also* heritage industrial

Jews 141, 237-44 *see also* Ghetto, Holocaust
jugendstijl see Art nouveau

Krüger (August) 212-13, 215

landscape 114, 157, 160, 163, 250
legal frameworks 4, 5, 7
legitimation 142, 248
local authorities 21, 37, 41, 77-82, 99, 121, 124, 155, 167, 180, 211-2, 251-2

marking 62, 89, 223-4, 239 *see also* heritage interpretation
memory 10, 243 see also amnesia
monument 38-9, 48, 91
   compensation 33, 34, 40
   decay 25, 105, 106, 120
   demolition 18, 22, 63, 65, 127, 139, 228
   criteria 20, 21, 257
   inventory 18, 32, 39, 191, 231-2, 257
   reconstruction 109-15, 140-1, 169-70, 229-31
   restoration 46, 48, 80, 90, 96-7, 107-13, 159
   reuse 51-2, 80-1, 82, 101, 117-130

selection 19, 20, 40, 247
statistics 20, 24, 40, 49, 59, 191-2, 231
types 48-9
Morris (William) 16
multicultural 5, 13, 89, 191 *see also* heritage 197, 256-7
slavery
museums 16, 33, 37, 62, 64, 68, 82, 92, 98, 121, 126, 135, 168, 191, 196, 197-8, 200-3, 207-21, 249 *see also* Skansen
culture 3, 215, 221
myths 29, 220

nation state 5
national
culture 6
governments 5
lottery 77, 80, 82, 125, 126, 154, 158, 160
trust 16, 18, 123, 124, 127
nationalism 36, 46, 52-3, 249-50
Natolini (Adolpho) 93-6
nature reserve 125

offices 80, 84, 97-8, 201-2
Oxenstierna (Axel) 30

Palladian 104, 216 *see also* classicism
parks *see* gardens
Pevsner (Nicholas) 17
picturesque 14
politics 52, 167, 194, 196-7
legitimation 52
place 254 *see also* space
identity 8, 58, 59-72, 87-100, 223-34
sense of 3, 7, 84-5, 163, 234, 237, 247
planning *see also* urban planning
Acts 19, 21, 26, 40-41, 233
participation 93-6
systems 37
postmodern 93, 99, 251
preservation *see* conservation, heritage, monuments
property 3, 18, 250

ownership 32, 64, 78, 101, 123-4, 154, 180, 250
rights 14, 15
public opinion 7
Pugin (Augustus) 15, 249

race 219, 220
relations 196-9
riots 196-7, 198
railway stations 96-9
regency 191-2
renaissance 213-14, 216-18
retailing *see* shopping
Ridley (Nicholas) 24
romanticism 14, 15, 34, 99, 204
Rijksdienst voor het Monumentenzorg 47, 52, 168
Ruskin (John) 15, 46, 249

Sandys (Duncan) 21
Scott (Giles Gilbert) 15, 46
Scott (Sir Walter) 14-15
Semper (Gottfried) 215-16, 217-18
shopping 51, 59, 82, 84, 93, 99, 135, 136, 162, 242, 252
signature buildings 88, 92
Skansen 35, 249
slavery 191-206, 256
Smythson (Robert) 119-20, 121
space 252 *see also* place
sacralisation 223, 224-5, 229
scale 8
time 67, 210-11
Stuers (Victor de) 46, 249
Stukeley (William) 14
sustainable 8, 125, 155, 158
symbols 9, 85, 87-90, 98, 182, 189-90, 201, 208-10, 223-34, 252, 254-7
coding 89, 203, 242

Telford (Thomas) 177
textualisation 60
Thomasz (Thomas) 134
tourism 133, 146, 237, 243, 252, 253
attraction 8, 101-2, 115-16, 123-6, 135-7, 189, 191, 196, 252-3
experience 141
facilities 124-5, 167, 168-9

grand tour 14
industry 156, 234
markets 142, 169, 170, 242, 243
policy 140-41, 156-7
spa 167-8, 170
visitors 115, 118, 121, 136-7, 141, 159, 184, 242
townscape 26, 88, 97-8
townschemes 79, 80

UNESCO 4, 104, 233 *see also* ICOMOS
urban
   design 91-2
   gentrification 25, 27, 52, 251-2
   growth 13
   imagery 88-90, 132, 163, 169, 224
   planning 10, 76, 79, 90-96, 171-2
   redevelopment 8, 253-4
   regeneration 8, 24, 57-100, 153-5, 160-61, 173, 192, 196-8, 202, 251-2
   renewal 26, 43, 62-5, 242-3
urbanisation *see* urban growth

values 5, 7, 60-61, 111
venacular 249
Victorian 19, 58, 73, 76, 122, 125
   Society 16, 77
Viollet-le-Duc (Eugene) 15, 46

Waern (Kolbjörn) 114-15
war *see* castles, fortress, world war two
water
   fronts 24
   lines 138
   table 138
World Heritage 9
   convention 104
   sites 227, 231, 232
World War Two 10, 13, 19, 27, 93, 237
Wornum (George) 230
Wren (Christopher) 19
Wyatt (James) 15
Wyatt (Jeffry) 120 also Wyattville

Yamasaki (Minoru) 9

Place Index

Africa 192-5, 200
Algeria 219
Australia 190

Bath (UK) 17, 23, 24, 79
Belgium 47
Birmingham (UK) 24
Bolsover (UK)146, 149-62, 254, 255
Bourtange (NL) 137-41, 169-70
Bradford (UK) 24, 155
Bristol (UK) 192, 196, 203
Britain *see* United Kingdom
Budapest (H) 239

Caribbean 192-3, 195
Chester (UK) 23
Chichester (UK) 23
China 193, 195
Coevorden (NL) 137, 140
Coventry (UK) 10

Denmark 6, 30, 248
Deventer (NL) 48
Dresden (G) 215, 217
Dubrovnik (Cr) 9

Egypt 219, 220
England 20-1, 11
Europe 4-5, 7, 145, 175, 219, 220, 237, 239, 242, 250
European Community *see* European Union
European Union 23, 29, 80, 108, 110, 116, 145, 154, 189

Finland 33
Forsvik (Sw) 146, 175-87, 254
France 29, 45, 47, 138
Friesland (NL) 138

Germany 29, 45, 141, 166, 167, 169, 170, 212, 215, 242
Glasgow (UK) 145, 155

Götaland (Sw) 30
Göta Canal (Sw) 177-8
Göteborg (Sw) 38, 58, 59-72, 103-16, 179, 212, 251
Gothenburg see Göteborg
Groningen (NL) 58, 87-100, 137, 138, 140, 166, 171, 237-44, 251-2, 256

India 193, 195
Israel *see* Palestine
Italy 47

Kalmar (Sw) 43
Karlsborg (Sw) 178, 182, 186
Kraków (P) 239

Leer (BRD) 166
Liverpool (UK) 153, 155, 189, 191-206, 256
London (UK) 15, 18, 21, 83, 124, 190, 192, 196, 223-34, 256
Londonderry / Derry (UK) 256
Louisbourg (Canada) 140

Maastricht (NL) 48
Malmö (Sw) 38
Manchester (UK) 20, 83
Middle East 241

Naarden (NL) 133-7
New York (USA) 9
Nieuweschans (NL) 137, 140, 146, 163-74, 254
North Africa 6, 241
Northern Ireland 16
Norrköpping (Sw) 38, 175

Nottingham (UK) 24, 58, 73-85, 117-30, 251, 255

Oudeschans (NL) 166

Palestine 219, 239
Poland 239, 242
Prague (Cz) 239

Scandinavia 6, 29, 169
Scotland 16, 18, 26, 140
Sheffield (UK) 153
Skansen (Sw) 35, 36
Stockholm (Sw) 33, 35, 42
Stonehenge (UK) 15
Sweden 6, 29-44, 59-72, 109, 190, 248-9

Tammerfors (Sw) 175
The Netherlands 6, 29, 45-54, 137-8, 248-9, 250

Theresienstad 238
Turkey 47, 242, 257

Uppsala (Sw) 34
United Kingdom 5-7, 13-28, 45, 47, 156-7, 161-2, 178, 248-9, 256
United States of America 9, 29, 196, 223, 239

Vänersborg (Sw) 207-21
Vilnius 239

Wales 16, 18, 26
Warsaw (P) 238-9
Willemstad (NL) 140
Winschoten (NL) 166, 167, 169

York (UK) 23
Yugoslavia 9